Lecture Notes in Computer Science

Vol. 142: Problems and Methodologies in Mathematical Software Production. Proceedings, 1980. Edited by P.C. Messina and A. Murli. VII, 271 pages. 1982.

Vol. 143: Operating Systems Engineering. Proceedings, 1980. Edited by M. Maekawa and L.A. Belady. VII, 465 pages. 1982.

Vol. 144: Computer Algebra. Proceedings, 1982. Edited by J. Calmet. XIV, 301 pages. 1982.

Vol. 145: Theoretical Computer Science. Proceedings, 1983. Edited by A.B. Cremers and H.P. Kriegel. X, 367 pages. 1982.

Vol. 146: Research and Development in Information Retrieval. Proceedings, 1982. Edited by G. Salton and H.-J. Schneider. IX, 311 pages. 1983.

Vol. 147: RIMS Symposia on Software Science and Engineering. Proceedings, 1982. Edited by E. Goto, I. Nakata, K. Furukawa, R. Nakajima, and A. Yonezawa. V. 232 pages. 1983.

Vol. 148: Logics of Programs and Their Applications. Proceedings, 1980. Edited by A. Salwicki. VI, 324 pages. 1983.

Vol. 149: Cryptography. Proceedings, 1982. Edited by T. Beth. VIII, 402 pages. 1983.

Vol. 150: Enduser Systems and Their Human Factors. Proceedings, 1983. Edited by A. Blaser and M. Zoeppritz. III, 138 pages. 1983.

Vol. 151: R. Piloty, M. Barbacci, D. Borrione, D. Dietmeyer, F. Hill, and P. Skelly, CONLAN Report. XII, 174 pages. 1983.

Vol. 152: Specification and Design of Software Systems. Proceedings, 1982. Edited by E. Knuth and E. J. Neuhold. V, 152 pages. 1983.

Vol. 153: Graph-Grammars and Their Application to Computer Science. Proceedings, 1982. Edited by H. Ehrig, M. Nagl, and G. Rozenberg. VII, 452 pages. 1983.

Vol. 154: Automata, Languages and Programming. Proceedings, 1983. Edited by J. Díaz. VIII, 734 pages. 1983.

Vol. 155: The Programming Language Ada. Reference Manual. Approved 17 February 1983. American National Standards Institute, Inc. ANSI/MIL-STD-1815A-1983. IX, 331 pages. 1983.

Vol. 156: M.H. Overmars, The Design of Dynamic Data Structures. VII, 181 pages. 1983.

Vol. 157: O. Østerby, Z. Zlatev, Direct Methods for Sparse Matrices. VIII, 127 pages. 1983.

Vol. 158: Foundations of Computation Theory. Proceedings, 1983. Edited by M. Karpinski, XI, 517 pages. 1983.

Vol. 159: CAAP'83. Proceedings, 1983. Edited by G. Ausiello and M. Protasi. VI, 416 pages. 1983.

Vol. 160: The IOTA Programming System. Edited by R. Nakajima and T. Yuasa. VII, 217 pages. 1983.

Vol. 161: DIANA, An Intermediate Language for Ada. Edited by G. Goos, W.A. Wulf, A. Evans, Jr. and K.J. Butler. VII, 201 pages. 1983.

Vol. 162: Computer Algebra. Proceedings, 1983. Edited by J.A. van Hulzen. XIII, 305 pages. 1983.

Vol. 163: VLSI Engineering. Proceedings. Edited by T.L. Kunii. VIII, 308 pages. 1984.

Vol. 164: Logics of Programs. Proceedings, 1983. Edited by E. Clarke and D. Kozen. VI, 528 pages. 1984.

Vol. 165: T.F. Coleman, Large Sparse Numerical Optimization. V, 105 pages. 1984.

Vol. 166: STACS 84. Symposium of Theoretical Aspects of Computer Science. Proceedings, 1984. Edited by M. Fontet and K. Mehlhorn. VI, 338 pages. 1984.

Vol. 167: International Symposium on Programming. Proceedings, 1984. Edited by C. Girault and M. Paul. VI, 262 pages. 1984.

Vol. 168: Methods and Tools for Computer Integrated Manufacturing. Edited by R. Dillmann and U. Rembold. XVI, 528 pages. 1984.

Vol. 169: Ch. Ronse, Feedback Shift Registers. II, 1–2, 145 pages. 1984.

Vol. 170: Seventh International Conference on Automated Deduction. Edited by R.E. Shostak. IV, 508 pages. 1984.

Vol. 171: Logic and Machines: Decision Problems and Complexity. Proceedings, 1983. Edited by E. Börger, G. Hasenjaeger and D. Rödding. VI, 456 pages. 1984.

Vol. 172: Automata, Languages and Programming. Proceedings, 1984. Edited by J. Paredaens. VIII, 527 pages. 1984.

Vol. 173: Semantics of Data Types. Proceedings, 1984. Edited by G. Kahn, D.B. MacQueen and G. Plotkin. VI, 391 pages. 1984.

Vol. 174: EUROSAM 84. Proceedings, 1984. Edited by J. Fitch. XI, 396 pages. 1984.

Vol. 175: A. Thayse, P-Functions and Boolean Matrix Factorization. VII, 248 pages. 1984.

Vol. 176: Mathematical Foundations of Computer Science 1984. Proceedings, 1984. Edited by M.P. Chytil and V. Koubek. XI, 581 pages. 1984.

Vol. 177: Programming Languages and Their Definition. Edited by C.B. Jones. XXXII, 254 pages. 1984.

Vol. 178: Readings on Cognitive Ergonomics – Mind and Computers. Proceedings, 1984. Edited by G.C. van der Veer, M.J. Tauber, T.R.G. Green and P. Gorny. VI, 269 pages. 1984.

Vol. 179: V. Pan, How to Multiply Matrices Faster. XI, 212 pages. 1984.

Vol. 180: Ada Software Tools Interfaces. Proceedings, 1983. Edited by P.J.L. Wallis. III, 164 pages. 1984.

Vol. 181: Foundations of Software Technology and Theoretical Computer Science. Proceedings, 1984. Edited by M. Joseph and R. Shyamasundar. VIII, 468 pages. 1984.

Vol. 182: STACS 85. 2nd Annual Symposium on Theoretical Aspects of Computer Science. Proceedings, 1985. Edited by K. Mehlhorn. VII, 374 pages. 1985.

Vol. 183: The Munich Project CIP. Volume I: The Wide Spectrum Language CIP-L. By the CIP Language Group. XI, 275 pages. 1985.

Vol. 184: Local Area Networks: An Advanced Course. Proceedings 1983. Edited by D. Hutchison, J. Mariani and D. Shepherd. VIII, 497 pages. 1985.

Vol. 185: Mathematical Foundations of Software Development. Proceedings, 1985. Volume 1: Colloquium on Trees in Algebra and Programming (CAAP'85). Edited by H. Ehrig, C. Floyd, M. Nivat and J. Thatcher. XIV, 418 pages. 1985.

Vol. 186: Formal Methods and Software Development. Proceedings 1985. Volume 2: Colloquium on Software Engineering (CSE). Edited by H. Ehrig, C. Floyd, M. Nivat and J. Thatcher. XIV, 455 pages. 1985.

Vol. 187: F.S. Chaghaghi, Time Series Package (TSPACK). III, 305 pages. 1985.

Vol. 188: Advances in Petri Nets 1984. Edited by G. Rozenberg with the cooperation of H. Genrich and G. Roucairol. VII, 467 pages. 1985.

Vol. 189: M.S. Sherman, Paragon. XI, 376 pages. 1985.

Vol. 190: M.W. Alford, J.P. Ansart, G. Hommel, L. Lamport, B. Liskov, G.P. Mullery and F.B. Schneider, Distributed Systems. Edited by M. Paul and H.J. Siegert. VI, 573 pages. 1985.

Vol. 191: H. Barringer, A Survey of Verification Techniques for Parallel Programs. VI, 115 pages. 1985.

Lecture Notes in
Computer Science

Edited by G. Goos and J. Hartmanis

235

Accurate Scientific Computations

Symposium, Bad Neuenahr, FRG, March 12–14, 1985
Proceedings

Edited by Willard L. Miranker and Richard A. Toupin

Springer-Verlag
Berlin Heidelberg New York London Paris Tokyo

Editors

Willard L. Miranker
Mathematical Sciences Department, IBM Research Center
Yorktown Heights, N.Y. 10598, USA

Richard A. Toupin
Department of Mechanical Engineering
Division of Applied Mechanics, Stanford University
Stanford, CA 94305, USA

CR Subject Classifications (1985): G.1, G.4, I.1

ISBN 3-540-16798-6 Springer-Verlag Berlin Heidelberg New York
ISBN 0-387-16798-6 Springer-Verlag New York Berlin Heidelberg

Library of Congress Cataloging-in-Publication Data. Accurate scientific computations. (Lecture notes in computer science; 235) 1. Mathematics–Data processing–Congresses. 2. Numerical calculations–Congresses. I. Miranker, Willard L. II. Toupin, Richard A., 1926-. III. Series.
QA76.95.A23 1986 510'.28'5 86-20364
ISBN 0-387-16798-6 (U.S.)

Printed in Germany

Printing and binding: Beltz Offsetdruck, Hemsbach/Bergstr.
2145/3140-543210

Accurate Scientific Computations

Preface

The theme of the symposium is the "accuracy" of certain kinds of mathematical results obtained with the aid of a computing machine. The symposium is entitled, "Accurate Scientific Computations", even though, as remarked by Stepleman in his introduction to Scientific Computing [1], "when different people use the words scientific computing, it is not only possible but probable that each has in mind something different."

No less than in mathematics, physics, chemistry, or any other branch of science, "scientific computing" cannot be defined independently of examples. This symposium brings together three quite different kinds of work, concepts of accuracy, and notions of scientific computation. A shared aspect of the work in the 12 papers presented at the symposium (9 of which are collected here), and its panel discussion, is the use of present day computing machines to address mathematical problems and questions. We are careful here to avoid using the term "numerical questions" so as not to exclude one of the three kinds of work represented in these papers; viz., Computer Algebra.

An alternative title for this symposium might be **Applications of Computing Machines in Mathematics.** Computing machines have come to be widely used as instruments of simulation and empiricism in what today is called "Scientific Computing". Important and useful as these applications of computers are in the various sciences and fields of engineering, they were not the dominant theme of this symposium. Rather it was algorithms which deliver precise results, both analytic and numerical. To express an indefinite integral of a rational function of elementary functions as a similar object, if and when it

---

1 North-Holland Publishing Co. (1983)

exists, is an example of the former. An algorithm which computes an error bound is an example of the latter. Another example of the latter is an algorithm which computes the probability that a real number lies in a prescribed interval. Some of the papers deal also with the efficiency of the implementations of such algorithms.

Scientific Computing has come to mean more narrowly the construction of solutions, or approximations of solutions of systems of differential or algebraic equations, or other constructive, finite, algorithmic processes of algebra and analysis. If we combine this narrower definition of "Scientific Computing" with the quest for mathematical theorems strictly proven, or computation automatically validated with the aid of a computing machine we arrive at the title of the symposium and a unifying concept for the results presented in the papers collected here. They address the idea of **Accurate Scientific Computation** in three quite different ways which we can illustrate with the important special and pervasive case of a "problem" in "Scientific Computing"; viz., "solving" a system of linear equations $Ax = B$.

To embrace all three concepts of accuracy in one simple and familiar example, we must narrow the problem even further and consider the case when the coefficient matrix A and the vector B are prescribed rational numbers with numerators and denominators of reasonable length. In this case, if the system is consistent, there exist rational solutions $x = (x^1, ..., x^n)$ and algorithms to compute each and every rational number x^i, $i = 1, 2, ..., n$. If the size of the system is not too large, it is a feasible task to compute and display the numerator and denominator of each and every x^i. A computer algebra "system" might implement such a display. It is one concept of an "accurate scientific computation". Of course, if the dimension of the system exceeds ten or twenty, then, in general, the numerators and denominators in this definition and representation of the "solution" may be very large integers indeed. The computation may be rather extensive and time consuming even on a large computer. But when A and B have small representations and the dimension of the linear system is small, there could be useful insight and purpose in this sort of accurate scientific computation. In particular, the precise integer rank of the matrix A could be determined in this way.

A second definition of the problem of "solving" the same system of linear equations $Ax = B$ is to construct (compute) a **floating-point** or other approximation $\tilde{x}$ to the rational solution x of the system if it be consistent, **and** to compute an upper bound on some norm of the difference $|x - \tilde{x}|$, **and** to require that this bound on the "error" of $\tilde{x}$ be less than some prescribed value. This approach is termed **validated computation.**

A third definition of the same "problem" is to compute a floating-point or other approximation $\tilde{x}$ to the same system of equations, **and** to compute (exhibit) a lower bound on the **probability** that the difference $|x - \tilde{x}|$ be not greater than some prescribed value.

Thus we have before us at least three quite different concepts of **Accurate Scientific Computing**, each of which is represented in the lectures and results collected here.

Basic to scientific computation is the evaluation of the elementary functions. In separate lectures by S. Gal and by F. Gustavson (abstract only) methods are described for computing very accurate values of the elementary scalar functions (sin, cos, log, square root, etc.) which, at the same time, are very fast. The speed and efficiency of the new algorithms exploit the architectural changes in computing machines which have occurred in the last two decades since the pioneering work of Kuki on this problem. Moreover, the new algorithms bound the relative error of the computed value of the function for all values of the argument. The bound guarantees the accuracy or significance of all but the last digit in the function value, and even the last for more than 99.9% of the arguments.

A review of concepts and results of a new and systematic theory of computer arithmetic is presented by U. Kulisch. The new arithmetic broadens the arithmetical base of numerical analysis considerably. In addition to the usual floating-point operations, the new arithmetic provides the arithmetic operations in the linear spaces and their interval correspondents, which are most commonly used in computation, with maximum accuracy on computers directly. This changes the interplay between computation and numerical analysis in a qualitative way. Floating-point arithmetic is defined concisely and axiomatically. The subsequent lectures by S. Rump, W. Ames, W. L. Miranker, and F. Stummel show aspects of this.

New computational methods to deal with the limitations inherent in floating-point arithmetic are presented by S. Rump. The mathematical basis is an inclusion theory, the assumptions of which can be verified by a digital computation. For this verification the new well-defined computer arithmetic is used. The algorithms based on the inclusion theory have the following properties:

- results are automatically verified to be correct, or when a rare exception occurs, an error message is delivered.
- the results are of high accuracy; the error of every component of the result is of the magnitude of the relative rounding error unit.
- the solution of the given problem is imputed to exist and to be unique within the computed error bounds.
- the computing time is of the same order as a comparable (purely) floating-point algorithm which does not provide these features.

The approach has thus far been developed for some standard problems of numerical analysis such as systems of linear and non-linear equations, eigenproblems, zeros of polynomials, and linear and convex programming. When data of a given problem is specified with tolerances, every problem included within the tolerances is solved and an inclusion of its solution is computed. The key property of the algorithms is that the error is controlled automatically. These concepts and this "validation" approach to scientific computation are collected in a subroutine library called ACRITH. The presentations of W. Ames and of W. Miranker develop other possibilities for the exploitation of ACRITH. In the latter presentation, methods for directly exploiting the new computer arithmetic are given as well.

W. Ames describes software for solving the finite difference equations corresponding to boundary value problems of elliptic partial differential equations.

The routines, programmed in VS Fortran, employ the ACRITH Subroutine Library, and provide the user a choice of any one of eleven classical algorithms for solving a system of linear equations. Each algorithm can be executed with traditional computer arithmetic, or with ACRITH. This permits the user to observe the advantages of using ACRITH. Illustrative data is presented.

W. Miranker shows that good arithmetic can improve algorithmic performance. Compared to results obtained with conventional floating-point arithmetic, the computations are either more accurate or, for a given accuracy, the algorithms converge in fewer steps to within the specified error tolerance. Two approaches are presented. First: the high performance linear system solver of ACRITH is used in the areas of regularization (harmonic continuation) and stiff ordinary differential equations. Second: the routine use of a highly accurate inner product (a basic constituent of the new floating-point arithmetic) is shown to result in acceleration of eigenelement calculations (QR-algorithm), the conjugate gradient algorithm and a separating hyperplane algorithm (pattern recognition). Not all algorithms are susceptable of improvment by such means and some speculations are offered.

Schauer and Toupin present a method for computing a bound on the error of an approximation to the solution of a restricted class of systems of linear equations. These systems include those arising from discretization of certain boundary-value problems of elliptic partial differential equations. They also present empirical evidence for the existence of a critical precision P(A) of floating-point arithmetic used in the conjugate gradient algorithm for constructing an approximation to the solution of a system (A) of linear equations. The critical precision P(A) depends on the system. If the precision of the floating-point arithmetic is less that P(A), then the residual fails to diminish monotonically as it would were the precision infinite. If the precision of the arithmetic exceeds P(A), they observe that the approximate residual diminishes montonically to zero in a number of "steps" of the algorithm not greater that the dimension of the system (were the precision infinite, the number of "steps" would be the number of distinct eigenvalues of the matrix (A)). Moreover, for each digit of precision in excess of P(A), one more significant digit in the approximate solution is obtained. For the large sparse systems investigated, the number of steps is a small fraction of the dimension if the precision is greater than P(A). The critical precision P(A) is an emprically determined "condition number" for a system of linear equations. It may be less than or greater than any of the precisions provided by the floating-point arithmetic units of a particular machine.

F. Stummel presents a new method for the derivation of exact representations of errors and residuals of the computed solutions of linear algebraic systems under data perturbations and rounding errors of floating-point arithmetic. These representations yield both strict and first-order optimal componentwise a posteriori error and residual bounds which can be computed numerically together with the solutions of the linear systems. Numerical examples of large linear systems arising in difference approximations of elliptic boundary value problems, in finite element and boundary integral methods show that the bounds so obtained constitute realistic measures of the possible maximal errors and residuals.

Some problems of algebra and analysis, such as obtaining explicit formulas for the derivative or integral of special classes of functions are finite computational tasks. One objective of "computer" algebra is to discover and implement such algorithms. Approximations to real numbers, such as provided by floating-point arithmetic, runs counter to the spirit of this work in computer algebra or "symbol manipulation". On the other hand, the results delivered by these algorithms, though finite, may be bewilderingly long. Taking integration as an example, J. Davenport shows how results of such computer algebra systems might be combined with numerical integration schemes to speed and enhance the accuracy of the latter.

The two lectures of B. Trager and G. E. Collins (abstracts only) also concern finite computational tasks in algebra for which no approximations to real numbers are invoked.

The problem of computing a bound on the error of an approximate solution of a system of linear equations, the value of an elementary function, or the root of a polynomial using only finite approximations to real numbers is not a trivial one, as placed in evidence by several of the papers presented at the symposium. What can one hope to do with the same question if applied to the floating-point approximations to solutions of large systems of non-linear equations in many variables such as those computed daily by the national and international weather bureaus? Indeed, it would seem a hopeless task if approached in the same spirit and with the same ideas that have been found effective for the elementary and fundamental sub-tasks of such large and complex computations involving billions of round-off errors. R. Alt and J. Vignes present an alternative question and means to address it. They replace the problem of computing error bounds on approximations by the problem of computing probabilistic estimates of the error of an approximation. In practice, their approach resembles the familiar scheme of computing two approximations with a large and complicated program; one using floating-point arithmetic with double the precision of the other. One gains some confidence, in this way, with the significance of common high-order digits in the two approximations. Alt and Vignes propose that one perturb the program and intermediate floating-point results in such large and complex computations in a systematic way, and infer the probability that the common leading digits of a small sample of approximations computed in this way are significant.

These lectures and the exchange of views of the panelists and participants during the panel discussion point to a continuing evolution and broadening of the concepts, objectives, and methods of Scientific Computation. The papers collected here provide evidence of the interplay between the discovery of algorithms for new and old mathematical tasks and the evolution of computer architectures. The theme of the work presented in these papers is "accuracy"; different concepts and definitions of it, ways to achieve it efficiently, and algorithms to prove or "validate" it. We foresee a gradual evolution of the objectives of Scientific Computing wherein the quest for "accuracy" competes in a more balanced way with the quest for "speed". We believe that the concepts, results, and methods described in the papers of this symposium will seed and influence such an evolution of the subject. In summary, these are:

- Efficient algorithms for evaluation of elementary functions having specified and guaranteed accuracy based on the non-standard Accurate Tables Method.

- Axioms for "computer arithmetic", including directed roundings (interval arithmetic).

- The theory of and techniques for computing inclusions.

- Computer architectures which implement essential primitives for achieving accuracy and proving it, such as high precision inner products, and variable precision floating-point arithmetic.

- Probabilisticalgorithms for estimating the accuracy of complex and extensive floating-point computations.

- A synergism of the concepts and methods of "computer" algebra (exact) computations, and those which invoke approximations to real numbers and functions of them.

Yorktown Heights, NY *W. L. Miranker*

Heidelberg, FRG *R. A. Toupin*

Acknowledgements

The symposium on **Accurate Scientific Computations** was held March 12-14, 1985 at the Steigenberger Kurhotel, Bad Neuenahr, Federal Republic of Germany.

It was sponsored and organized by IBM Deutschland GmbH Scientific Programs, Prof. Dr. C. Hackl, with the assistance of Ms. E. Rohm.

The Scientific Chairman was Dr. S. Winograd, Director Mathematical Sciences Department, IBM Research Center, Yorktown Heights, N.Y. Dr. H. Bleher, Dr. W. L. Miranker, and Dr. R. A. Toupin were associate organizers. The sessions were chaired by Dr. S. Winograd, Prof. Dr. L. Collatz, Prof. Dr. R. Loos, and Dr. A. Blaser.

There was a panel discussion chaired by Prof. Dr. P. Henrici. The panel members were Prof. F. W. J. Olver, Prof. Dr. H. J. Stetter, and Prof. Dr. H. Werner.

Table of Contents

COMPUTING ELEMENTARY FUNCTIONS : A NEW APPROACH FOR ACHIEVING HIGH ACCURACY AND GOOD PERFORMANCE

SHMUEL GAL

IBM Israel Scientific Center
Haifa, Israel

ABSTRACT

We present a method developed in the IBM Israel Scientific Center for designing algorithms for computing the elementary mathematical functions. This method which we call the "Accurate Tables Method" achieves high performance and produces very accurate results.

Our method is based on a table lookup and then a minimax polynomial approximation of the function near the table value. It overcomes one of the main problems encountered in elementary mathematical functions computations of achieving last bit accuracy even for the double precision routines. This task is difficult since using extended precision calculations (or simulating them) leads to a significant degradation of the performance. We found a way to obtain correctly rounded results for more than 99.9% of the argument values without using extended precision calculations.

Our main idea in the Accurate Tables Method is to use "nonstandard tables" different from the natural tables of equally spaced points in which the rounding error prevents obtaining last bit accuracy. In order to achieve a small error we use the following idea: Perturb the original, equally spaced, points in such a way that the table value (or tables values in case we need several tables) will be very close to numbers which can be exactly represented by the computer (much closer than the the usual double precision representation). Thus, we were able to control the error introduced by the computer representation of real numbers and extended the accuracy without actually using extended precision arithmetic.

1. INTRODUCTION

The current VS Fortran library of the elementary functions mainly consist of subroutines written by Kuki in the early 60'S for the IBM System/360. The IBM machines developed for the 1980's possess entirely new computational characteristics along with improved microcode features that make it profitable to revise these subroutines in order to take advantage of the capabilities of the new machines. This paper presents a methodology designed for the new machines.

We assume that the machines to be considered can use a table lookup during the computation without significantly degrading the performance. (Usually one table lookup, possibly extracting two or three neighbouring entries, is sufficient).

For convenience of presentation we assume that IBM 370 floating point arithmetic is used. In some implementations, such as in the 4300 series microcode, one can take advantage of the possiblity to obtain the result of the multiplication of two double precision floating point numbers (with 56-bit mantissas each) in a 64-bit register , and also to add (or subtract) two such registers. However, our results are not limited to any specific implementation.

In addition to the above assumption we also take into account the following property possesed by the new machines:
Floating point multiplication is fast while division is relatively slow.

In this paper we present a method, to be denoted as the Accurate Tables Method, for designing algorithms which have high performance and produce very accurate results. Later, we illustrate the use of such a method by presenting a detailed description of such an algorithm for computing the logarithm function.

2. THE GENERAL COMPUTATIONAL SCHEME.

The existing subroutines for calculating an elementary function f(x) are based on two steps. The first step is range reduction, i.e., transforming the original task into calculating f(x) (or a closely related function) in a specific predetermined interval A <= x < B. Some such examples for possible range reduction are listed below:

(1) sin(x)=sin(y) (or -sin(y)), where 0<=y<PI and y=x-n*PI.

(2) e**x =(2**n)*(2**(-y))
where 0<=y<1 and n=int(x/ln(2))+1.

(3) ln(x)=n*ln(2)+ln(y)
where .5<=y<1 and n is the binary exponent of x.

(4) sqrt(x)=(2**n)*sqrt(y)

where $.25 <= y < 1$ and $x = (4^{**}n)^*y$.

The above described step usually requires only a small effort. The second step is usually a rational (sometimes polynomial) approximation for f(x) in (A,B). One exception is the use of Newton-Raphson iterations for calculating the square root.

In the Accurate Tables Method we also use a range reduction, somewhat similar to the standard one, but the second step is quite different. A priori, we determine a set S of L points:

(5) $X1 < X2 <.. < XL$ in the interval (A,B) where
$X(i+1)-Xi <= d$, d being a predetermined constant.

In order to calculate f(x) for x in (A,B) we find a point Xi in S which is "close enough" to x and calculate f(x) using a table value (or some table values) associated with f(Xi), and a polynomial in x-Xi. To clarify the idea we present two such examples:

A. In order to calculate sin(y) (or cos(y)) where y lies in (0, pi/4) one can find an Xi with $-d/2 <= y-Xi < d/2$ and use the identity

(6) $\sin(y) = \sin(Xi)\cos(h) + \cos(Xi)\sin(h)$, where $h = y-Xi$.

The values of sin(Xi) and cos(Xi) can be extracted from a preprepared table , while cos(h) and sin(h) can be approximated by minimax polynomials in (-d/2 , d/2) .

B. In order to calculate 2**-y where $0 <= y < 1$ one can use the identity

(7) $2^{**}-y = (2^{**}-Xi)^*(2^{**}-h)$, where $h = y-Xi < d$.

The value of (2**-Xi) can be extracted from a preprepared table, while (2**-h) can be approximated by a minimax polynomial p(x) in (0,d).

The distance d is determined as a compromise between the two contradictory goals: 1) As d gets smaller so does the degree of the minimax polynomial p(x) needed to achieve the required accuracy. Thus, the subroutine performance (speed) is improved. Working with a small d has the additional advantage from the accuracy point of view. (The accuracy problem will be discussed in Section 3). On the other hand: 2) The overall size of the tables is limited by the space allocated to them in the computer memory and thus d cannot get be too small. A practical value for d used by us is usually 1/256 .

A general purpose program for determining the minimax polynomial for any given function is presented in <4>.

3. THE ACCURACY PROBLEM.

An important goal of our algorithms is achieving last bit accuracy. This means that if the correct result (containing an infinite number of bits) is rounded so as to produce a floating point number with a mantissa Q , then a desired goal is to have all the bits of the result R produced by our algorithm to coincide with those of Q. Achieving last bit accuracy for all argument values of the functions of the elementary mathematical library seems a very difficult task since these functions are transcendental (except for the square root where the Tuckerman test <3> can be used). Thus, any finite calculation of such functions involves a certain truncation error which may, in certain instances, lead to an incorrect rounding of the result R.
As an example, assume that our subroutine produces results in double precision and that the truncation error is 2**-65. Now, if bits 57 - 65 of the correct result are 011111111 then bits 57 - 65 of our result would be 100000000 which would lead to an incorrect rounding of bit 56 of R. Thus, the correctness of the last bit is very difficult to be guaranteed.

The accuracy problem is not severe for single precision calculations because a fast double precision arithmetic can be used. Since the relative truncation error can be quite easily controlled (say below 2-**40) it follows that the single precision algorithms can yield last bit accuracy which is close to 100% without significantly affecting the performance. On the other hand, achieving last bit accuracy in the double precision subroutines is difficult since using extended precision calculations (or simulating them) leads to a significant degradation of the performance. Our Accurate Tables Method overcomes this difficulty and achieves last bit accuracy for more than 99.9% argument values without using extended precision. This method was so designed that the achievement of very high accuracy did not lead to any significant degradation of the performance.

In order to describe our method we first analyze the sources of numerical errors induced by the computational scheme described in Section 2. There are three factors contributing to this error:

T1. Using tables consisting of double precision floating point numbers induces a (relative) rounding error which may reach 2**-56 for a mantissa with its first bit equal to 1. Since IBM computers use base 16 arithmetic it follows that there can be 3 leading zeros in the mantissa which may induce a (relative) rounding error of 2-**53. In the next section we will present a new idea for a drastic reduction of T1.

T2. The truncation error of the minimax polynomial. This error can be quite easily reduced in most cases from (say) 2-**58 to 2-**66 by adding one extra term in the minimax polynomial.

T3. The error associated with the double precision arithmetic which uses a 56 bit mantissa. This sort of error can be reduced by a proper implementation of the algorithm. For example, in computing 2**-y , 0<y<1, needed in our algorith for computing exp(x), we use

(8) 2**-y = (2**-Xi)*(2**-h) , where h=y-Xi.

The value of 2**-Xi can be obtained by a table value Fi and 2**-h can be approximated by

(9) P(h)= 1+C1*h+.. +Cn*h**n =
=((..(Cn*h+C(n-1))*h+.. +C1)*h+1= P1(h) +1 .

Now, if !h!< 2**-8 then

(10) !P1(h)!< .7*2**-8 .

Thus instead of calculating (8) by Fi*P(h) we can obtain a more accurate result by calculating P1(h) and then computing (8) as

(11) Fi*P1(h) + Fi .

If Fi*P1(h) is obtained in 64-bit mantissa and the addition is carried in 64 bits, then the relative error of (11) is less than 2**-65 (using symmetric rounding). Actually, we can achieve a small error even without using a 64-bit mantissa. In fact, if we work with 56 bits (using the IBM 370 double precision arithmetic) and add a rounding term of magnitude 2**-57 to Fi*P1(h) before adding it to Fi in Expression (11) we would obtain a high accuracy. In this case the error in the last bit of Fi*P1(h) which could be generated in that (double precision) multiplication would be shifted over bit 64 of (11) because of inequality (10).

In order to estimate the effect of the errors on the last bit accuracy consider, for example, the above calculation and assume that the maximal error induced by the table values, the polynomial approximation and the arithmetic is smaller than 2**-65 with an average magnitude of 2**-66. Now, if the error is +2**-66 then our result will be incorrectly rounded if and only if bits 57-66 of the mantissa of the correct result are 0111111111 , while if the error is -2**-66 then the corresponding configuration is 1000000000 . Since the probability of each such configuration is about 1/2**10 (we will deal with the probabilities of bit configurations in the next section) it follows that the probability of an incorrectly rounded result is about 1/1000 so that the frequency of correctly rounded results is about 99.9%.

We have thus seen that errors of type T2 and T3 can be quite easily controlled to stay below a predetermined limit (e.g. 2**-65). The next section presents a method for controlling T1.

4. A METHOD FOR OBTAINING HIGHLY ACCURATE TABLES

As already mentioned in Section 2, our scheme for computing f(x) in (A,B) involves using table values of f, and possibly some other related functions, at a predetermined set of points S. The natural choice for such a set is equally spaced points i.e., Xi = A + i*d , i= 0,1,.. ,L-1. If we use this set, however, the rounding error of the tables, denoted by T1 in Section 3, will affect the last bit (sometimes even several bits) of our result. We now show that by choosing Xi in a specific "nonstandard" manner it is possible to drastically reduce T1.

To illustrate our idea we present a simple example in which only one value Fi =f(Xi) is needed. In order to achieve a small error, say less than 2**-65, we use the following idea: Try to perturb the original set S by choosing

(12) Xi = A + i*d + Ei

where Ei are small numbers, say less than d/1024 , and Xi posseses the following property:

(13) Bits 57-65 of the mantissa of f(Xi) are all 0 or all 1.

Condition (13) can be easily checked by an extended precision calculation of f(Xi). The choice 0< =Ei< = d/1024 (or a similar requirment) follows from the fact that Ei has to be small enough so as not to reduce the accuracy of the minimax polynomial approximation in any noticeable way.

Finding a number Xi which satisfies property (13) is not difficult because of the following reason : Consider the interval

(14) A+i*d <= x < A+i*d +d/1024

If one chooses a random number uniformly distributed in (14), then, except for the trivial case f(x)=constant, it will be shown at the end of this section that

(15) The probability that (13) holds is (about) 2**-8

Now, the number K of floating point double precision numbers in the interval (14) is

(16) K= d/(1024*w)

where w is the difference between two consecutive double precision machine numbers. For example, in the IBM 370 floating point double precision arithmetic if (14) is a subinterval of (1/16,1) then w=2**-56 , while if (14) is a subinterval of (1,16) then w=2**-52. Thus for all subintervals of (-16,16), which covers all the numbers Xi used by any of our algorithms, w <= 2**-52. Thus, if d=2**-8 then

(17) K >= 2**34 .

Thus, by property (15) the expected number of Xi in the interval (14) which satisfy property (13) is at least 2**26. Obviously, we need only one such number. This number can sometimes be found by an analytic method using the specific structure of f(x), but a practical method that will work for any function is just to carry a random (uniform) sampling in the interval (14) until such a number is found. By (15), the expected sample size is 2**8 = 256.

The above idea can be extended to more complicated situations in which the needed table consists of several entries. For example, in computing ln(x) (to be discussed in detail in the next section) or in computing sin(x), we use two entries of the table : f(Xi) and g(Xi). In such a situation we look for numbers Xi in the interval (14) which satisfy

(18) Bits 57-65 of both mantissas of f(Xi) and g(Xi)
are all 0 or all 1.

The existence of such numbers follows from the fact to be discussed at the end of this chapter that if f(x) and g(x) are "independent" functions, then the following property holds:

(19) The probability that (18) holds is (about) 2**-16
(2**-8 * 2**-8) .

Thus, by (17), the expected number of Xi satisfying (18) in the interval (14) is at least 2**18. Once again, a feasible method for obtaining Xi is random sampling in the interval (14) with an expected sample size of 2**16. It should be noted that the task of finding Xi of the set S has to be done only once (when preparing the tables).

The important feature of property (18) is the reduction of the rounding error of the tables by a factor of 2**-9 (or even more if a stricter requirement is imposed instead of (18)). Using this idea we are able to replace the relative error of our computations below 2**-65. As mentioned before, reducing the error of the result even to that extent does not always guarantee last bit accuracy. However, both theoretical considerations and practical experience show that such an error yields last bit accuracy for more than 99.9% of the argument values. In the next section we demonstrate in detail how to obtain high accuracy in computing ln(x). A similar idea can be used for an accurate computation of the trigonometric and other elementary functions.

Before we continue to the next section we would like to demonstrate the validity of statements (15) and (19).This can be shown as follows: Divide the interval given by (14) into N subintervals (Qn,Qn +h), n=1,.. ,N such that in each subinterval both f and g can be approximated by the following linear functions

(20) f= C*x+u+o(2**-65) and
g= D*x+v+o(2**-65)

where C=C(n) and D=D(n) are two different irrational numbers and 0< =x< =h (These conditions hold if we choose, for example, h=2**-40).

Now, assume for convenience of presentation that $1/16 <= x,f,g < 1$. Since x is a double precision floating point number, then (using x as the original argument minus Qn)

(21) x= j*2**-56   j=1,2,.. ,J

where J is a large number, and f and g can be represented as fractions to the base 256, i.e.

(22) f= .L1 L2 ..   L7 L8 ... and
    g= .M1 M2 ..   M7 M8 ...

where $0 <= Li,Mi <= 255$.

(In the more general case x=x1*16**K , f=f1*16**K1 , g=g1*16**K2 , where $1/16 <=$ x1,f1,g1 < 1 and K,K1,K2 are integers, the constants appearing in expression (20) : C, u, D, and v are replaced, respectively, by C*16**(K-K1), u*16**-K1, D*16**(K-K2), and v*16**-K2). Thus, it follows from (20),(21) and (22) that

(23) j*C+U+o(2**-8)= f*2**56= L1 L2 .. L7.L8 ... and
    j*D+V+o(2**-8)= g*2**56= M1 M2 .. M7.M8 ...

where U= u*2**56, V= v*2**56, and the right side of (23) is a number to the base 256. Since C and D are two different irrational numbers it follows from a well known theorem in Number Theory that the fractional parts of j*C and j*D , j=1,2, ... , are uniformly distributed in the unit square (see Cassels <1> Chapter IV). This implies that the fractional parts of j*C+U and j*D+V are also uniformly distributed in the unit square. Thus, by (23), the fractional parts of f*2**56 and g*2**56 are "almost" uniformly distributed in the unit square; this means that for all $0 <= k,m <= 255$

(24) Pr(L8=k and M8=m)=(about) (2**-8)*(2**-8)= 2**-16.

Since the above statement holds for all subintervals Qn <= x <= Qn+h, it holds for the whole interval given by (14). Statement (24) and the above discussion imply (19) and (15).

The above proof holds if the integer j appearing in (21) is unbounded. In our case j is bounded by a large number J. (For example, if h= 2**-40 then J= 2**-40 /2**-56 = 2**16). Thus, our results about the probabilities in (15) and (19) are only approximate. Practical experiments carried out by us while preparing the tables for various elementary functions indicate that (15) and (19) indeed hold for all practical purposes. (It is possible to find some pathological intervals in which the derivatives of f and g are very close to one another so that (19) does not hold for that interval but this happens in very rare cases).

5. AN ACCURATE AND EFFICIENT ALGORITHM FOR COMPUTING ln(X).

We now illustrate the use of the Accurate Tables Method by describing in detail an algorithm for computing the natural logarithm function. This algorithm is important not only for computing the logarithm itself but also for obtaining an accurate result for x**y using our ln(x) and exp(x) algorithms (with some minor modifications). Thus we have a large improvement not only in the ln(x) but also in the POWER subroutine which may be prone to a substantial error under the existing Fortran subroutine (see <2> page 84).

A. Mathematical Description

Write x=y*2**n where n is an integer and .75< =y<1.5. (We use this range reduction rather than the one described in (3) in order to avoid cancellation errors for x less than and close to 1).
Then ln(x)=ln(y)+n*ln2. To calculate ln(y) we use a table of triplets Xi , ln(Xi) , 1/Xi , where for i=0 to 192 : Xi= .75+ i/256 +1/512 +Ei (Ei are small numbers to be discussed later).
Then ln(y) = ln(Xi*y/Xi) = ln(Xi) + ln(1+(y-Xi)*1/Xi) = ln(Xi)+ln(1+z) where z = (y-Xi)*1/Xi. Since Xi can be chosen very close to y we have in all cases -1/384 < z < 1/384 and if y is close to 1 then (about) -1/512 < z < 1/512. We use a polynomial approximation p(z) to obtain ln(1+z) (degree 3 for single and 6 for double precision).

If x is near 1 then we use a minimax polynomial approximating ln(x) without using the table lookup. The relative error of the approximation is less than 2**-31 for single and less than 2**-72 for double precision.

The table Xi, Fi,Gi i=0 to 191: (It is a table of triplets of double precision words. Thus it contains 576 numbers).
Fi=ln(Xi) where Xi is defined above and Gi=1/Xi .
The numbers Xi were chosen by us in such a way that although Fi and Gi contain only 56 bit mantissas they have a relative accuracy of 2**-65Ü This is done by searching near .75+i/256+1/512 for such a number Xi such that bits 57 - 67 of the mantissas of ln(Xi) and of 1/Xi will be all 0 or all 1. (That is why the small numbers Ei were introduced). Both Fi and Gi are double precision words obtained by an extended precision calculation and symmetric rounding. This table is used for both the single and the double precision routines.

B. Details of the Algorithm:

Start. If x <= 0: error. Otherwise write x as x = (16**exp)*(2**-k)*y where .75 <= y < 1.5 (binary normalization by left shifts of the mantissa of x).
Define: n = 4*exp-k (then ln(x) = ln(y) + n*ln2).

(a) If the binary representation of y is y = .11xxxxxx u (u represents the bits from 9 and up) then extract the integer i given by bits 3 - 8 (6 bits) 0 <= i <= 63.

For single precision take z = (y-Xi)*Gi and calculate p = z + z**2*(D2 + z*1/3) (or ((z*1/3 + D2)*z + 1)*z if that is quicker) where D2 = -.5-3*2**-22, and 1/3 are given as double precision numbers.

For double precision distinguish the following two cases:

ad1. If i < 62 take z = (y-Xi)*Gi and calculate
p = ((((z*B6 + B5)*z + B4)*z + B3)*z + B2)*z**2 + z
where the hexadecimal representation of B1-B6 is listed below:

B2 = c0800000 00000000

B3 = 40555555 55542292

B4 = c03fffff fffb43b8

B5 = 40333340 ed2e5464

B6 = c02aaac3 6a5b04cc .

(The coefficients of p were obtained by our genral purpose program for minimax polynomial approximation using the restriction that the leading coefficient is zero and that the linear coefficient is 1.)
A method for calculating p by only 5 multiplications (instead of 6) with the same number of additions as before is:

A = B3/B5

B = B5/B6

C= B4/B6 -B3/B5

D=B2/B6 -A*C

zz=z*z

p= zz*B6*((zz+A)*(zz+B*z+C)+D) +z .

ad2. Otherwise do not calculate z as before but take z = y-1 and calculate:

p=((((((D6*z+D5)*z+D4)*z+D3)*z+D2)*z+D1)*z+D0)*z**2+z

where D0 -D6 are double precision numbers given as follows:

(Calculate from extended precision and round symmetrically)

D0= -.5

D1=1/3+4.63589337e-15

D2= -.25-12*2**-50

D3= .2-4.609092511e-10

D4= -1/6+43.2*2**-36

D5=1/7+1.302690578938e-5

D6= -.125-57.6*2**-22 .

A method for calculating p which needs only 6 multiplications (instead of 8) with the same number of additions as before is:

Ti=Di/D6 i=1 TO 5

A1=1.423743587533184

A2=T5

A3=T4-A1

A4=T2/T5-A3*A1/T5

A5=T0-A4*(T3-A3*T5-A4)

A6= T3-A3*T5-A4-A1*A2

Now calculate p as follows:

zz=z*z

p= D6*zz*(((zz+A3)*z+A4)*((zz+A1)*(z+A2)+A6)+A5)+z

(b) If y=1.0xxxxxxx u, extract the integer j which consists of bits 3 - 9 (7 bits) 0 < = j < = 127 and put i=j+64.

b1. For single precision: calculate p as in (a).

For double precision :

bd1. If j> =2, p is calculated as in ad1 with
z = (y-Xi)*Gi .

bd2. Otherwise, p is calculated as in ad2 with
z = y-1. In all cases, except ad2 and bd2, add p to Fi. Denote the result by r.

The final step is to calculate n*ln2 where ln2 is represented by one double precision number H1 for the single precision algorithm and as a sum of two numbers H1 and H2 for the double precision algorithm (H1 =40b1 7217 0000 0000 and H2=3af7 d1cf 79ab c9e4 (in hexadecimal representation)). Then add the result to r.

In order to avoid roundoff errors it is best to carry the arithmetical operations as follows:

((...   )*z**2
+n*H2
+z
+(Fi +n*H1).

C. Performance and Accuracy

It should be noted that the performance and accuracy of the above algorithm depend on the specific implementation. A correct implementation of the double precision algorithm on an IBM 370 machine yields last bit accuracy for more than 99.9% of the argument values of, practically, any interval. (For a single precision routine our last bit accuracy is virtually 100%).

In contrast, in a test we made with an argument uniformly distributed in (1.,2.) we found that VS Fortran subroutine has last bit accuracy of about 5% and also produces an error of two bits or more in more than 50%. For the interval (1.5,1.7) the last bit accuracy of

VS Fortran is 0% and the frequency of an error of two bits or more is 99%. (The last bit accuracy of the single precision VS Fortran routine for these intervals is about 70%).

Our algorithm should also yield a significantly faster routine. (Our microcode implementation is about twice faster than the routines used by VS Fortran).

ACKNOWLEDGMENT

The author would like to express his gratitude to Alex Hauber, David Jones and Israel Tzur for their excellent programming and implementation effort.

REFERENCES

<1> J. W. S. Cassels, An Introduction to Diophantine approximation, Cambridge University Press 1957.

<2> W. Cody and W. Waite , Software Manual for the Elementary Functions, Prentice-Hall 1980.

<3> B. Tuckermam, IBM T. J. Watson Research Center, Private Communication. The test is described in P. 14 of IBM Elementary Math Library, Program Reference and Operations Manual, SH20-2230-1, Second Edition (August 1984).

<4> I. Tzur and S. Gal, A General Purpose Program for Minimax Polynomial Approximation, IBM Israel Scientific Center Classified Technical Report.

International Scientific Symposium of IBM Germany
"ACCURATE SCIENTIFIC COMPUTATIONS"
March 12-14, 1985, Bad Neuenahr

FAST ELEMENTARY FUNCTION ALGORITHMS FOR 370 MACHINES

Dr. F. G. Gustavson, IBM Research Yorktown Heights

Abstract:

We describe new fast and accurate elementary function algorithms. The functions looked at, both in long and short precision were sqrt. x, z!, e^x, ln x, log x, sin x, cos x, tan x, cotan x, arctan x, arctan y/x, arcsin x, arcos x, and x^y. These algorithms were implemented in System 370 assembler language. These new routines have nearly perfect accuracy. In fact, some of the new algorithms always provide the correctly rounded result while in all cases we were able to guarantee that at worst it is the last bit which is in error. The latter routines return the correctly rounded result about 95 % of the time. This should be contrasted with the existing elementary function codes in which, at times, several digits are in error. Surprisingly, the new routines are also faster than the existing elementary function codes, sometimes by as much as a factor of two. We shall present both accuracy and speed comparisons for several of our functions. We will also describe our basic algorithmic approach of table look-up and briefly show why we can simultaneously achieve both high accuracy and speed.

A NEW ARITHMETIC FOR SCIENTIFIC COMPUTATION

Ulrich Kulisch
Institut für Angewandte Mathematik
Universität Karlsruhe
Postfach 6380, D-7500 Karlsruhe

Summary: The paper summarizes an extensive research activity in computer arithmetic and scientific computation that went on during the last fifteen years. We also discuss the experience gained through various implementations of a new approach to arithmetic on diverse processors including microprocessors.

We begin with a complete listing of the spaces that occur in numerical computations. This leads to a new and general definition of computer arithmetic.

Then we discuss aspects of traditional computer arithmetic such as the definition of the basic arithmetic operations, the definition of the operations in product spaces and some consequences of these defintions for error analysis of numerical algorithms.

In contrast to this we then give the new defintion of computer arithmetic. The arithmetic operations are defined by a general mapping principle which is called a semimorphism. We discuss the properties of semimorphisms, show briefly how they can be obtained and mention the most important feartures of their implementation on computers.

Then we show that the new operations can not be properly addressed by existing programming languages. Correcting this limitation led to extensions of PASCAL and FORTRAN.

A demonstration of a computer that has been systematically equipped with the new arithmetic will be given. The new arithmetic turns out to be a key property for an automatic error control in numerical analysis. By means of a large number of examples we show that guaranteed bounds for the solution with maximum accuracy can be obtained. The computer even proves the existence and uniqueness of the solution within the calculated bounds. If there is no unique solution (e.g. in case of a singular matrix) the computer recognizes it. Toward the end of the paper we sketch how expressions or program parts can be evaluated with high accuracy.

A. The Spaces of Numerical Computation

In addition to the integers, the real numbers **R**, the complex numbers **C**, the real and complex intervals **IR** and **IC** as well as the vectors and matrices over all of these: **VR**, **MR**, VC, **MC**, VIR, **MIR**, VIC and **MIC** comprise the fundamental numerical data types in computation. We present a table in Fig. 1 in the second column of which the spaces just enumerated are found. Since generic elements in these spaces are not representable in computers, each is replaced by a computer representable subset as listet in the third column of Fig. 1. In that column R denotes the computer representable counterpart of **R**, C the set of all pairs of elements of R, VC the set of all n-tuples of such pairs, and so forth.

The powerset of any set S is denoted by PS. The powerset of several sets of column 2 are displayed in column 1.

We indicate set-subset relations between elements of neighboring columns in Fig. 1 by means of the inclusion symbol $\supset$.

Having described the sets listed in the third column of Fig. 1, we turn to the arithmetic operations to be defined in these sets. These operations are supposed to approximate the operations in the corresponding sets listed in the second column. In general the latter are not computer executable even for computer representable operands. The operations required for each of the computer representable sets listed in column 3 are in turn indicated in column 4.

Moreover, a number of outer multiplications (e.g. matrix-vector or a scalar matrix multiplication) are required, and these in turn are indicated in column 4 by means of a x-sign between rows of the figure. By computer arithmetic, we understand all operations that have to be defined in all of these sets listed in the third column of Figure 1 as well as in certain combinations of these sets. In a good programming system, these operations should be available as operators for all admissible combinations of these sets with maximum accuracy.

1		2		3	4
		R	⊃	R	+ − × /
					x
		VR	⊃	VR	+ −
					x
		MR	⊃	MR	+ − ×
PR	⊃	IR	⊃	IR	+ − × /
					x
PVR	⊃	IVR	⊃	IVR	+ −
					x
PMR	⊃	IMR	⊃	IMR	+ − ×
		C	⊃	C	+ − × /
					x
		VC	⊃	VC	+ −
					x
		MC	⊃	MC	+ − ×
PC	⊃	IC	⊃	IC	+ − × /
					x
PVC	⊃	IVC	⊃	IVC	+ −
					x
PMC	⊃	IMC	⊃	IMC	+ − ×

Figure 1. Table of the spaces occuring in numerical computations

B. Traditional Definition of Computer Arithmetic

From the many operations that occur in the column under R in Fig. 1 traditional Computers in general provide only four, the addition, subtraction, multiplication and division of floating-point numbers in R. All the others have to be defined by the user himself in case he needs them. He can only do so by means of subroutines. Each occurance of an operation in an algorithm then causes a procedure call. This is a complicated, time consuming and numerically unprecise approach.

It turns out that there are in principle two different basic methods for defining computer arithmetic. In the sense of Fig. 1, these methods may be called the vertical and the horizontal method. By way of illustration of these two possibilities, we consider in Fig. 2, a simple detail of Fig. 1. In Fig. 2 we display the sets **R**, R, **C**, C and **MC** and MC. On traditional computers a floating-point arithmetic is usually available. We indicate this by an arrow drawn from **R** to R in Fig. 2. By the vertical definition of the arithmetic in C and MC. We mean that the operations in C and MC are defined by the operations in R and the usual formulas for the addition, multiplication and division of complex numbers respectively the addition and multiplication of complex matrices. On most computers this is precisely the method of definition of the arithmetic operations in product spaces. All operations in the sets of the column under R are defined by the given operations in R.

$$\begin{array}{ccc} \mathbf{R} & \longrightarrow & R \\ \downarrow & & \downarrow \\ \mathbf{C} & & C \\ \downarrow & & \downarrow \\ \mathbf{MC} & & MC \end{array}$$

Figure 2

It is well known that this method causes a complicated error analysis in the sets of the column under R.

C. The New Definition of Computer Arithmetic by Semimorphisms

The horizontal method of defining the arithmetic operations in the sets under R in Fig. 1 is a process which passes from left to right by using a certain set of formulas. We begin with the definition of the operations in the powersets listed in column 1 of Fig. 1. Let M be one of the sets **R**, **VR**, **MR**, **C**, **VC** and **MC**. Then the operations in the powerset PM are defined by

$$A * B := \{a * b \mid a \in A \wedge b \in B\}$$

for each $* \in \{+,-,\times,/\}$ and for all pairs of sets $A,B \in PM$. With this definition the arithmetic operastions are well defined in the leftmost element in each row of Fig. 1. Now let M be any set whatever in Fig. 1 in which the operations are well defined and let N be the subset of M occuring immediately to its right in the figure. For each operation $*$ in M a corresponding operation $\boxast$ in N is defined as follows:

(RG) $\quad x \boxast y := \square(x * y)$ for all $x,y \in N$.

Here $\square : M \rightarrow N$ denotes a projection, which we call a rounding, from M into N. This mapping has the following properties:

(R1) $\square x = x$ for all $x \in N$ (rounding)

(R2) $x \leq y$ implies $\square x \leq \square y$ for all $x, y \in M$ (monotonicity)

(R4) $\square(-x) = -\square x$ for all $x \in M$ (antisymmetry)

In the case that M is a set of intervals $\leq$ denotes set inclusion $\subseteq$ and the rounding has the additional property

(R3) $x \subseteq \square x$ for all $x \in M$ (upwardly directed)

With the obvious modification of these rules in the case of outer operations, this completes our review of the horizontal definition of computer arithmetic.

The above formulas show that the mapping $\square$ is close to being a homomorphism. (R1) is a quite natural property, which every rounding should have. (RG), (R2) and (R3) can be shown to be necessary conditions for a homomorphism between ordered algebraic structures. Therefore, we call the mapping which they characterize a semimorphism.

It is important to understand that the operations defined by semimorphism in general are different from those defined by the vertical method which we discussed proviously.

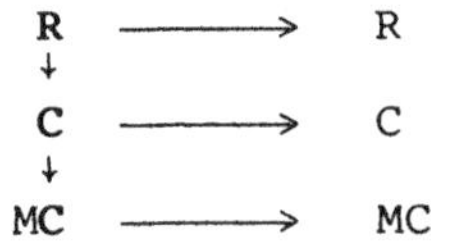

Figure 3

The horizontal method defines the operations in a subset N of a set M directly by making use of the operations in M. So a semimorphism directly links an operation and its approximation in the subset. It is easy to see now that a semimorphism of a semimorphism is a semimorphism again. With this disovery operations by semimorphisms in all sets of Figure 1 are directly defined.

The new operations defined for all sets of Fig. 1 deliver maximum accuracy in the sense that between the correct result of an operation and its approximation in the subset is no further element of the subset /6/, /7/. This fundamental result is guaranteed by (RG), (R1) and (R2).

To conveniently perform all these operations in the case of interval spaces, we introduce the monotone directed roundings ∇ respectively Δ from **R** into R which are defined by (R1), (R2) and

(R3) $\nabla x \leq x$ resp. $x \leq \Delta x$ for all $x \in \mathbf{R}$.

Along with these roundings, we shall make use of the associated operations defined as follows:

(RG) $x \mathbin{\nabla\!\!\!\!*} y := \nabla(x * y)$ resp. $x \mathbin{\Delta\!\!\!\!*} y := \Delta(x * y)$
for all $x,y \in R$ and $* \in \{+,-,\times,/\}$.

D. Implementation of Semimorphisms on Computers

We now descuss the question whether in all cases of Fig. 1 the operations defined by semimorphism can be implemented on computers by means of fast algorithms. At first sight it seems doubtful that formula (RG) in particular can be implemented on computers at all. In order to determine the approximation $x \boxast y$, the correct result $x * y$ seems to be necessary. If $x * y$ is representable on the computer we need not to approximate it. In general $x * y$ will not be representable and not executable on the computer. In this case it seems that it can not be used to define $x \boxast y$. Especially according to definition (RG), all intervall operations are ultimately based on operations in the powerset which are not executable in a computer.

Therefore we seek to express the operations defined in the sets of column 3 of Fig. 1 in terms of executable formulas. This can be done by means of isomorphisms. It can not be the place here to develop these ideas in all details. For a detailed discussion see /6/, /7/. We just mention that for a verification of these isomorphisms the weak structures that hold in the subsets of column 4 in Fig. 1 are to be used. These subsets must be analysed for this purpose.

In /6/ and /7/ it is shown that all semimorphic operations in the sets of the third column of Fig. 1 can be realized on computers by a modular technique in terms of a higher programming language if

1. an operator concept or an operator notation is available for all operations in that higher level language and

2. the following 15 fundamental operations for floating-point numbers are available

$\boxplus$	$\boxminus$	$\boxtimes$	$\boxslash$	$\boxdot$
$\overset{+}{\nabla}$	$\overset{-}{\nabla}$	$\overset{\times}{\nabla}$	$\overset{/}{\nabla}$	$\overset{\cdot}{\nabla}$
$\overset{+}{\Delta}$	$\overset{-}{\Delta}$	$\overset{\times}{\Delta}$	$\overset{/}{\Delta}$	$\overset{\cdot}{\Delta}$

Here $\boxast$, $* \in \{+,-,\times,/\}$, denotes the semimorphic operations defined by (RG) using one monotone and antisymmetric rounding (R1,2,3) as for instance a rounding to the nearest number of the screen. $\overset{*}{\nabla}$ and $\overset{*}{\Delta}$, $* \in \{+,-,\times,/\}$, denote the operations defined by (RG) and the monotone downwardly respectively upwardly directed rounding. $\boxdot$, $\overset{\cdot}{\nabla}$ and $\overset{\cdot}{\Delta}$ denote scalar products with maximum accuracy

$$\circ\left(\sum_{i=1}^{n} a_i \times b_i\right), \quad \circ \in \{\Box, \nabla, \Delta\} .$$

Algorithms for the implementation of these operations can be found in /2/, /6/, /7/. Of these 15 fundamental operations traditional numerical analysis only makes use of the four operations $\boxplus$, $\boxminus$, $\boxtimes$, $\boxslash$. Traditional interval arithmetic uses the eight operations $\overset{+}{\nabla}$, $\overset{-}{\nabla}$, $\overset{\times}{\nabla}$, $\overset{/}{\nabla}$ and $\overset{+}{\Delta}$, $\overset{-}{\Delta}$, $\overset{\times}{\Delta}$, $\overset{/}{\Delta}$. The newly proposed IEEE Computer Society Arithmetic offers twelve of these operations $\boxast$, $\overset{*}{\nabla}$, $\overset{*}{\Delta}$, $* \in \{+,-,\times,/\}$. These twelve operations were systematically implemented in software on a Zuse Z23 in 1967 and in hardware on a X8 in 1968 at the Institute for Applied Mathematics at the University of Karlsruhe. Both implementations were supported by a high level language "TRIPLEX ALGOL-60", published in Numerische Mathematik in 1968, /1/.

The historic developement has shown that many difficulties that traditionally occur in numerical analysis as well as in interval analysis cannot be avoided unless the three scalar products $\boxdot$, $\overset{\cdot}{\nabla}$ and $\overset{\cdot}{\Delta}$ are available on the computer for all relevant values of n. Algorithms providing the three scalar products can and should be made available on every computer by fast hardware routines. We note in passing, that optimal scalar products can be provided on a computer by a black box technique, where the vector components a_i and b_i, i = 1(1)n, are the input and the scalar products the output.

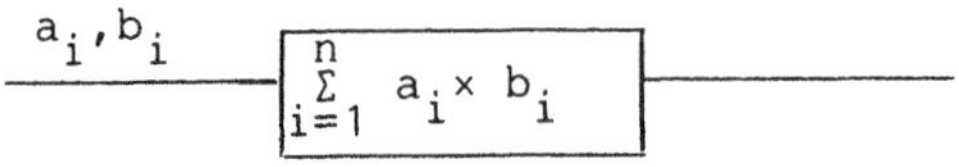

$$c = \circ \sum_{i=1}^{n} a_i * b_i$$

$$\circ \in \{\Box, \nabla, \Delta\} .$$

The black box only needs some local storage and works independently of the main storage of the computer. The size of the local storage depends only on the data formats in use (base of the number system, length of the mantissa and range of the exponents). In particular it is independent of the dimension n of the two vectors $a = (a_i)$ and $b = (b_i)$ to be multiplied. Scalar products play a key role for an automatic error analysis and the computation of small bounds for the solution of the problem by the computer itself. If realized in hardware they lead to a considerable gain in speed whenever scalar products occur in a computation.

A complete implementation of the arithmetic operations defined by semimorphisms in all spaces of Fig. 1 was first executed at the author's Institute between 1977 and 1979. PASCAL is used as the basic language. It was extended by a general operator concept. The new arithmetic has a tremendeous impact on numerical analysis. Program packages are available for all kinds of standard problems such as linear systems of equations, inversion of matrices, eigenvalue and eigenvector problems, zeros of polynomials, nonlinear systems of equations, evaluation of arithmetic expressions, linear and quadratic optimization problems, numerical quadrature, initial and boundary value problems of ordinary differential equations, iterative solution of large systems of equations, and so on. The computer is equipped with a twelve decimal digit arithmetic. For all problems just cited results can generally be guaranteed to at least 11 digits. The computation time is in all cases comparable with corresponding programs using ordinary floating-point arithmetic.

The new techniques do more than provide sharp bounds for the solution. They permit the computer to simultaneously verify the existence and the uniqueness of the solution within the computed bounds. If there is no unique solution (for instance in the case of a singular matrix) the computer detects this fact and delivers this information to the user.

We are now going to sketch these techniques briefly. With optimal scalar products the step from the execution of the single arithmetic operations +, -, *, and / with maximum accuracy to the computation of the value of rational arithmetic expressions (mathematical functions) with maximum accuracy can be made.

As a first example let us consider the following scalar product of matrices:

$$A * B + C * D + E * F \qquad (*)$$

If all operations that occur in this expression are defined by semimorphisms they are of maximum accuracy. Evaluation of the expression on computer, however, in general does not deliver maximum accuracy. This is the case because of a possible accumulation of rounding errors. However since scalar products with maximum accuracy are already available the whole expression (*) can be computed with maximum accuracy and only one rounding (componentwise) at the end.

We demand, therefore, that the programming language in use allows the evaluation of those vector and matrix expressions with maximum accuracy that can be evaluated with maximum accuracy by scalar products. These expressions are sums. As summands (terms) may occur real or complex scalars, vectors, matrices, and products of these.

In the programming language such "scalar product expressions" may be written in parenthesis with a rounding symbol in front of them. As rounding symbols may be used $\square$, ∇ , Δ and $\Diamond$. For instance

$\square$(A * x - b) or ∇(I - R * A)

Here $\square$ means the implicit rounding of the computer which should be a monotone and antisymmetric rounding. ∇ and Δ senote the monotone directed roundings. $\Diamond$ means that the expression has to be rounded to the least including interval by ∇ and Δ .

Scalar product expressions are a real breakthrough in Numerical Analysis. The meaning may become clear by considering the two matrix expressions:

A * B + C * D + E * F and $\square$(A * B + C * D + E * F)

In the first expression each sum may cause cancellation. Since the products are rounded this may be tragic. In the second expression cancellation is harmless, because the full information is kept until the final rounding.

Sclar product expressions with maximum accuracy allow an effective use of defect correction methods in Numerical Analysis. This means that information that has already been lost during a computation often finally can be recovered.

The following two talks will show - among others - that linear systems of equations very generally can be solved with maximum accuracy by use of optimal scalar products. With a similar technique rational expressions can be evaluated with maximum accuracy.

As a model situation, consider the following polynomial of degree three.

$$p(t)=a_3t^3+a_2t^2+a_1t+a_0=((a_3t+a_2)t+a_1)t+a_0 ,$$

where a_0,a_1,a_2,a_3 and t are given floating-point numbers. The expression on the right hand side is called the Horner scheme. Evaluation of p(t) by means of the Horner scheme procedes in the following steps:

$$\begin{array}{llllll}
x_1=a_3 & & x_1 & & & =a_3 \\
x_2=x_1t+a_2 & & -tx_1 + x_2 & & & =a_2 \\
x_3=x_2t+a_1 & \text{or} & & -tx_2 + x_3 & & =a_1 \\
x_4=x_3t+a_0 & & & & -tx_3 + x_4 & =a_0
\end{array}$$

This is a system of linear equation Ax = b with a lower triangular matrix, where

$$A=\begin{pmatrix} 1 & 0 & 0 & 0 \\ -t & 1 & 0 & 0 \\ 0 & -t & 1 & 0 \\ 0 & 0 & -t & 1 \end{pmatrix}, x=\begin{pmatrix} x_1 \\ x_2 \\ x_3 \\ x_4 \end{pmatrix} \text{ and } b=\begin{pmatrix} a_2 \\ a_2 \\ a_1 \\ a_0 \end{pmatrix} .$$

x_4 is the value of the polynomial. Then a maximally accurate solution of the linear system delivers the value of the polynomial with maximum accuracy. The extension to higher order polynomials is obvious. This procedure generates maximally accurate evaluation of polynomial, even of very high order.

Let us now consider general arithmetic expressions and begin with the example

$$(a + b)c - d / e .$$

Here a,b,c,d and e are floating-point numbers. Evaluation of this expression can be performed in the following steps.

$$\begin{aligned}
x_1 &= a \\
x_2 &= x_1 + b \\
x_3 &= cx_2 \\
x_4 &= d \\
ex_5 &= x_4 \\
x_6 &= x_3 - x_5 .
\end{aligned}$$

Once again we obtain a linear system of equations with a lower triangular matrix.

There are arithmetic expressions which lead to a non-linear system of equations. For example, the expression

$$(a + b)(c + d)$$

leads to the non-linear system of equations

$$\begin{aligned} x_1 &= a \\ x_2 &= x_1 + b \\ x_3 &= c \\ x_4 &= x_3 + d \\ x_5 &= x_2 x_4 \ . \end{aligned}$$

All such systems are of a special lower triangular form. They can be transferred into linear systems by an automatic algebraic transformation process /13/. Solution techniques which employ optimal scalar products and defect correction methos can then be used. In this way the value of the arithmetic expression with maximum accuracy is obtained. The extension of computation with maximum accuracy from dyadic operations (even in the product spaces of Fig. 1) to arbitrary arithmetic expressions is fundamental. Even though the operations are implemented optimally, in computations involving several such operations errors may accumulate and become large. With optimal scalar products and defect correction methods we can reduce the loss of information in the evaluation of polyadic operations or arithmetic expressions to only one single rounding.

In many cases it is more user friendly to express the computation of expressions by means of a conventional program part. For example, suppose the expressions are already so encoded. The user desiring to upgrade results from such a piece of code so that they are maximally accurate is not obliged to re-program. He may just upgrade his program.

For example, let **PROG** stand for the statement sequence of such a program part. Let x,y,z be names of those variables whose values are computed within **PROG** which are to be upgraded to outputs with maximum accuracy. This is accomplished in the following way

```
accurate x<,y>,z do
   begin
      PROG
   end
```

This modified program computes x,y,z with maximum accuracy and rounds x downwardly, y upwardly, z to the nearest.

Needless to say, accurate evaluation of expressions or program parts is slower than execution with simple floating-point. However, accurate evaluation obviates the need for an error analysis. It also may be critical in unstable cases. In case of a polynomial, for instance, a direct-evaluation of the polynomial in floating-point may be unstable. The corresponding linear system is stable.

The technique for expression evaluation, which we just described, can also be applied in case of matrix or vector expressions. In this case the matrix of the resulting linear system is a sparse block matrix of lower triangular form.

Expression evaluation with maximum accuracy, as described in this paper, differs essentially from traditional expression evaluation. The expression or program part to evaluate with maximum accuracy is first transformed into a non linear or linear system of equations by a complex rewrite system. The latter is than evaluated in floating-point making use of optimal scalar products. The rewrite system consists of symbolic manipulations only. It should be part of the compiler.

Floating-point has been invented for high speed computations. A disadvantage is that it often leads to inaccurate results. This drawback is avoided by techniques like symbolic manipulation, seminumerical algorithms or rational arithmetic. The latter on the other hand are often slow and require the use of a different programming environment. The methodology for accurate floating-point evaluation of expressions, suggested in this paper, combines the high performence quality of floating-point with the safety which is provided with these other methods. It allows the user to stay in his customary programming environment.

References:

/1/ Apostolatos, N., Kulisch, U., Krawczyk, R., Lortz, B., Nickel, K., Wippermann, H.-W.: The Algorithmic Language TRIPLEX ALGOL 60, Num. Math. 11, 175-180 (1968)

/2/ Bohlender, G.: Floating-point Computation of functions with maximum accuracy. IEEE Trans. Comp. C-26, Nr. 7, 621-632 (1977)

/3/ Bohlender, G., Kaucher, E., Klatte, R., Kulisch, U., Miranker, W.L., Ullrich, Ch. und Wolff von Gudenberg, J.: FORTRAN for contemporary numerical computation, Report RC 8348, IBM Thomas J. Watson Research Center 1980 and Computing 26, 277-314 (1981)

/4/ Kulisch, U.: An axiomatic approach to rounded computations, TS Report No. 1020. Mathematics Research Center, University of Wisconsin, Madison, Wisconsin, 1969 und Numer. Math. 19, 1-17 (1971)

/5/ Kulisch, U.: Interval arithmetic over completely ordered ringoids, TS Report No. 1105, Mathematics Research Center, University of Wisconsin, Madison, Wisconsin, 1970

/6/ Kulisch, U.: Grundlagen des Numerischen Rechnens - Mathematische Begründung der Rechnerarithmetik. Reihe Informatik, Band 19, Wissenschaftsverlag des Bibliographischen Instituts Mannheim, 1976

/7/ Kulisch, U., Miranker, W.L.: Computer Arithmetic in Theory and Practice, Academic Press, 1980

/8/ Coonan, J. et al.: A proposed standard for floating-point arithmetic, SIGNUM newsletter, Oct. 1979

/9/ INTEL 12 1586-001: The 8086 family user's manual, Numeric Supplement, July 1980

/10/ Kulisch, U., Miranker, W.L. (Editors): A New Approach to Scientific Computation, Academic Press, 1983

/11/ Kulisch, U., Miranker, W.L.: The Arithmetic of the Digital Computer, IBM Research Report RC 10580, 1984, to appear in SIAM Reviews

/12/ High Accuracy Arithmetic, Subroutine Library, IBM Program Description and User's Guide, Program Number 5664-185, 1984

/13/ Böhm, H.: Berechnung von Polynomnullstellen und Auswertung arithmetischer Ausdrücke mit garantierter maximaler Genauigkeit. Dissertation, Universität Karlsruhe 1984

Additional References are given in /7/, /10/ and /11/

New Results on Verified Inclusions

Siegfried M. Rump

Abstract. The computational results of traditional numerical algorithms on computers are usually good approximations to the solution of a given problem. However, no verification is provided for some bound on the maximum relative error of the approximation. As can be demonstrated by ill-conditioned examples, those approximations may be drastically wrong. The algorithms based on the inclusion theory (cf. [38]) do have an automatic verification process. Rather than approximations to the solution an inclusion of the solution is computed and the correctness of the bounds together with the existence and uniqueness of the solution within the bounds is automatically verified by the computer without any effort on the part of the user. The computing time of these algorithms is of the order of a comparable, standard floating-point algorithm (such as Gaussian elimination in case of general linear systems).

In the following some new results complementing the inclusion theory are given. One of the main results is that the inclusion sets need not to be convex. Therefore other types of inclusion sets such as torus-sectors can be used. Another main observation is that the new and old theorems can be proved without using fixed point theorems. Moreover improvements of existing theorems of the inclusion theory by means of weaker assumptions are presented.
Another fundamental observation is the following. It is well-known that a real iteration in $\mathbb{R}^n$ with affine iteration function converges if and only if the spectral radius of the iteration matrix is less than one. It can be shown that a similar result holds for our inclusion algorithm: An inclusion will be achieved if and only if the spectral radius of the iteration matrix is less then one. This result is best possible.

It is demonstrated by means of theorems and examples that even for extremely ill-conditioned examples very sharp inclusions of the solution are computed. The inclusions are almost always of least significant bit accuracy, i.e. the left and right bounds of the inclusion are adjacent floating-point numbers.

1. Introduction. Let $\mathbb{R}$ be the set of real numbers, $V\mathbb{R}$ the set of real vectors with n components, $M\mathbb{R}$ the set of real square matrices with n rows, $\mathbb{C}$ the set of complex numbers, $V\mathbb{C}$ the set of complex vectors with n components and $M\mathbb{C}$ the set of complex square matrices with n rows. In the following the letter n is reserved to denote the number of elements of a vector or the number of rows and columns of a square matrix. Vectors with other than n components are, for instance, denoted by $V_{n+1}\mathbb{R}$, non-square matrices for example with ℓ rows and m columns over the complex numbers by $M_{\ell,m}\mathbb{C}$. I denotes the identity matrix.

The operations in the power set PT over T for $T \in \{\mathbb{R}, V\mathbb{R}, M\mathbb{R}, \mathbb{C}, V\mathbb{C}, M\mathbb{C}\}$ are as usual defined by

$$A \in PT,\ B \in PT: \quad A * B := \{a * b \mid a \in A,\ b \in B\} \tag{1}$$

for $* \in \{+,-,\cdot,/\}$ and well-known restrictions for /. Definition (1) applies for inner and outer operations. In case a set (an element of the power set) occurs more than once in a formula special care has to be taken. Consider as an example $f : PV\mathbb{R} \to PV\mathbb{R}$ being defined by $f(x) := Z + \mathfrak{C}\cdot X$ for $X, Z \in PV\mathbb{R}$ and $\mathfrak{C} \in PM\mathbb{R}$. Then

$$f(f(X)) = Z_1 + \mathfrak{C}_1 \cdot (Z_2 + \mathfrak{C}_2 \cdot X) \quad \text{for} \quad Z_1 = Z_2 = Z \quad \text{and} \quad \mathfrak{C}_1 = \mathfrak{C}_2 = \mathfrak{C}.$$

Moreover

$$\{z + C \cdot (z + C \cdot x) \mid z \in Z,\ C \in \mathfrak{C},\ x \in X\} \subseteq f(f(x)) \tag{2}$$

where in general eguality does not hold in (2).

The order relation $\leq$ in $\mathbb{R}$ is extended to $V\mathbb{R}$ and $M\mathbb{R}$ by

$$A,B \in V\mathbb{R}: \quad A \leq B \ :\Leftrightarrow\ A_i \leq B_i \quad \text{for} \quad 1 \leq i \leq n \quad \text{and}$$

$$A,B \in M\mathbb{R}: \quad A \leq B \ :\Leftrightarrow\ A_{ij} \leq B_{ij} \quad \text{for} \quad 1 \leq i,j \leq n.$$

The order relation in $\mathbb{C}$ is defined by

$$a + bi,\ c + di \in \mathbb{C}: \quad a + bi \leq c + di \ :\Leftrightarrow\ a \leq c \quad b \leq d$$

and similarly in VC and MC.

The set of intervals $\mathbb{I}T$ over T for $T \in \{\mathbb{R}, V\mathbb{R}, M\mathbb{R}, \mathbb{C}, VC, MC\}$ is defined by

$$[A,B] \in \mathbb{I}T :\Leftrightarrow [A,B] := \{x \in T \mid A \le x \le B\} \quad \text{for } A,B \in T \text{ and } A \le B.$$

Therefore $\mathbb{I}T \subseteq PT$ where every element of $\mathbb{I}T$ is nonempty , convex, closed and bounded.

By S we denote some finite set of real numbers (which could be regarded as the set of single precision floating-point numbers on some computer). We consider the set of n-tupels VS over S and the set of n^2 - tupels MS over S. Similarly for some finite subset $\mathbb{C}S$ of $\mathbb{R}\times\mathbb{R}$ we consider $V\mathbb{C}S$ and $M\mathbb{C}S$. If U is some set out of S, VS, MS, $\mathbb{C}S$, $V\mathbb{C}S$, $M\mathbb{C}S$ and T is the corresponding set in $\mathbb{R}$, $V\mathbb{R}$, $M\mathbb{R}$, $\mathbb{C}$, $V\mathbb{C}$, $M\mathbb{C}$, then intervals over U are defined by

$$[A,B] \in \mathbb{I}U :\Leftrightarrow [A,B] := \{x \in T \quad A \le x \le B\} \quad \text{for } A,B \in U \text{ and } A \le B .$$

For example, an interval $[a,b] \in S$ consists of all real numbers x with $a \le x \le b$. Usually, interval operations $\diamond\!\!\!\!*$, $* \in \{+, -, \cdot, /\}$ are defined by

$$A,B \in \mathbb{I}T : \quad A \diamond\!\!\!\!* B := \cap\{M \in \mathbb{I}T \mid A * B \in M\} \tag{3}$$

and

$$A,B \in \mathbb{I}U : \quad A \diamond\!\!\!\!* B := \cap\{M \in \mathbb{I}U \mid A * B \in M\} \tag{4}$$

for T and U as above. Using (3) and (4) the operations $\diamond\!\!\!\!*$ are well defined (cf. [4],[28]). Sometimes the strict definition (4) is difficult to realize on computers, e.g. for complex division. Therefore we allow in our further discussions any isotone interval operation $\circledast$, i.e. any operation

$$\circledast : PT \times PT \to \mathbb{I}T \quad \text{with } A,B \in PT \Rightarrow A * B \subseteq A \circledast B \text{ respectively} \tag{5}$$

$$\circledast : PU \times PU \to \mathbb{I}U \quad \text{with } A,B \in PU \Rightarrow A * B \subseteq A \circledast B . \tag{6}$$

There will no confusion be caused by the fact that in $\mathbb{I}T$ and $\mathbb{I}U$ the same symbol $\circledast$ is used. In (5) and (6) elements of the power set over T or U are allowed to define more general interval operations. As follows by (3) and (4), $\diamond\!\!\!\!*$ is best possible for interval arguments. We allow any interval rounding $\lozenge : PT \to \mathbb{I}T$ resp. $\lozenge : PU \to \mathbb{I}U$ satisfying

$$A \in PT : \quad A \subseteq \lozenge A \quad \text{and } A \in PU : \quad A \subseteq \lozenge A .$$

Here $O : PT \to \mathbb{I}T$ resp. $O : PU \to \mathbb{I}U$ has to be regarded as an operator rather than a mapping. That means we do not necessarily require

$$OA = OB \quad \text{for} \quad A = B.$$

Instead the essential property for O is $A \subseteq OA$ for $A \in PT$ resp. $A \in PU$. The reason for this is that in practical implementations of O it might happen that $OA \neq OB$ for $A = B$ because of performance reasons. An overestimation of A may be allowed to save computing time, but this does not affect the following theorems.

There is always a best rounding $\Diamond$ (cf.[4]) satisfying

$$A \in PT : \Diamond A = \cap\{B \in \mathbb{I}T : A \subseteq B\} \quad \text{and}$$

$$A \in PU : \Diamond A = \cap\{B \in \mathbb{I}U : A \subseteq B\} .$$

The rounding $\Diamond$ and the operations $\circledast$ over $\mathbb{I}U$ for $* \in \{+,-,\cdot,/\}$ and $U \in \{S, VS, MS, CS, VCS, MCS\}$ can be effectively implemented on computers (cf.[25]).

For an interval $X = [A,B]$ the lower and upper bound is defined by $\inf(X) := A$ and $\sup(X) := B$, or, in short notation $\underline{X} := A$ and $\overline{X} := B$ (where is no confusion with the algebraic closure).

In the following we use beside the ordinary inclusion $\subseteq$ three other types of inclusions:

$$A,B \in P\,V\mathbb{R} : \quad A \overset{\circ}{\subseteq} B :\Leftrightarrow A \subseteq \overset{\circ}{B},$$

$$A,B \in \mathbb{I}\,V\mathbb{R} : \quad A \subsetneqq B :\Leftrightarrow A \subseteq B \;.\wedge.\; A_i \neq B_i \text{ for all } i,\ 1 \le i \le n,$$

$$A,B \in \mathbb{I}\,V\mathbb{R} : \quad A \tilde{\subseteq} B :\Leftrightarrow A \subseteq B \;.\wedge.\; A_i \neq B_i \text{ for some } i,\ 1 \le i \le n.$$

The definitions apply similarly over S. It is

$$A,B \in \mathbb{I}\,V\mathbb{R} : \; A \overset{\circ}{\subseteq} B \Rightarrow A \subsetneqq B \Rightarrow A \tilde{\subseteq} B.$$

For any algorithm performing verification of the results on a computer a precisely defined computer arithmetic is mandatory. Preferably this computer arithmetic should deliver highly accurate or maximum accurate results. An arithmetic delivering always results of least significant bit accuracy even for matrix and vector operations has been given in [8].

Of course, composed operations need not to be of high accuracy. Such highly accurate results for composed operations and even for whole algorithms for solving linear or nonlinear systems are delivered by the inclusion theory as described in the following.

2. Basic new results. In [41] one of the main steps to develop the inclusion theory was to apply Brouwer's Fixed Point Theorem to an affine mapping and to show contraction in order to obtain a zero from the fixed point. The following theorem gives Brouwer's Fixed Point Theorem for affine mappings, the elegant proof of which the author learnt at the University of Karlsruhe.

Theorem 1. Let X be a nonempty, compact, convex subset of the normed vector space E, $C : E \to E$ a linear transformation with continuous restriction $C/_X$, $z \in E$ and $T(X) \subseteq X$ for the affine mapping $Tx := z + C \cdot x$. Then there is a fixed point $\hat{x} \in X$ of $T : T\hat{x} = \hat{x}$.

Proof. Let $x^o \in X$ and define

$$x^m := \frac{1}{m+1} \cdot \{x^o + Tx^o + \ldots + T^m x^o\}, \quad m \geq 1.$$

Then short computation shows

$$Tx^m = \frac{1}{m+1} \cdot \{Tx^o + T^2 x^o + \ldots + T^{m+1} x^o\}$$

and by the convexity of X follows $Tx^m \in X$. Therefore

$$Tx^m - x^m = \frac{1}{m+1} \cdot \{T^{m+1} x^o - x^o\}$$

and because X is bounded $Tx^m - x^m \to 0$ for $m \to \infty$. Therefore some subsequence (x^{m_n}) of (x^m) converges to some $\hat{x} \in X$ with $T\hat{x} = \hat{x}$. □

To come from a fixed point of the affine mapping to a zero of the original problem (e.g. a system of linear or nonlinear equations) in [37] the contraction of the linear part is shown by assuming a proper inclusion $TX \overset{o}{\subseteq} X$ rather than $TX \subseteq X$. Next we will show this contraction for general compact subsets of $\mathbb{R}^n$ without assuming convexity.

Lemma 2. Let $Z \in \mathbb{P}V\mathbb{R}$, $\mathfrak{C} \in \mathbb{P}M\mathbb{R}$ and $\emptyset \neq X \in \mathbb{P}V\mathbb{R}$ compact. Then

$$\begin{aligned} &a) \quad Y = ch(X) \text{ and } Z + \mathfrak{C} \cdot X \overset{o}{\subseteq} Y \quad \Rightarrow \quad Z + \mathfrak{C} \cdot Y \overset{o}{\subseteq} Y. \\ &b) \quad Y = X - X \text{ and } Z + \mathfrak{C} \cdot X \overset{o}{\subseteq} X \quad \Rightarrow \quad \mathfrak{C} \cdot Y \overset{o}{\subseteq} Y. \end{aligned} \qquad (7)$$

Remark. All operations are the power set operations, ch(X) denotes the convex hull.

Proof. ad a): Let $z \in Z$ and $C \in \mathfrak{C}$. Then $y \in Y \Leftrightarrow \exists x_1, x_2 \in X : y = x_1 + \lambda(x_2 - x_1)$ for $0 \le \lambda \le 1$. Then with $x_3 := z + C \cdot x_1$ and $x_4 := z + C \cdot x_2$ follows $z + C \cdot y = x_3 + \lambda(x_4 - x_3) \overset{\circ}{\subseteq} Y$ by $x_3, x_4 \overset{\circ}{\in} Y$ and (7).

ad b): For $z \in Z$, $C \in \mathfrak{C}$ is $y \in X - X \Leftrightarrow \exists x_1, x_2 \in X : y = x_2 - x_1$. Therefore $C \cdot y = (z + C \cdot x_1) - (z + C \cdot x_2) \overset{\circ}{\subseteq} Y$. □

The proof in [40] of assertion b) in lemma 2 for convex, compact sets required Brouwer's Fixed Point Theorem by $z + C \cdot \hat{x} = \hat{x} \Rightarrow C \cdot (X - \hat{x}) = C \cdot X + z - \hat{x} \subseteq X - \hat{x}$. The assumption of proper inclusion $\overset{\circ}{\subseteq}$ can be sharpened for interval vectors. For this purpose we need the following lemma.

Lemma 3. Let $Z \in \mathbb{P}V\mathbb{R}$, $\mathfrak{C} \in \mathbb{P}M\mathbb{R}$ and $X \in \mathbb{I}V\mathbb{R}$. Then for $Y := X \ominus X$ holds

a) $\quad \lozenge(Z + \mathfrak{C} \cdot X) \overset{\circ}{\subseteq} X \quad \Rightarrow \quad \lozenge(\mathfrak{C} \cdot Y) \overset{\circ}{\subseteq} Y$

b) $\quad \lozenge(Z + \mathfrak{C} \cdot X) \subsetneqq X \quad \Rightarrow \quad \lozenge(\mathfrak{C} \cdot Y) \overset{\circ}{\subseteq} Y$

c) $\quad \lozenge(Z + \mathfrak{C} \cdot X) \tilde{\subseteq} X \quad \Rightarrow \quad \lozenge(\mathfrak{C} \cdot Y) \tilde{\subseteq} Y$.

Proof. ad a) and b): Suppose $\lozenge(Z + \mathfrak{C} \cdot X) \subsetneqq X$ and $z \in Z$, $C \in \mathfrak{C}$. Obviously $Y = [\underline{X} - \overline{X}, \overline{X} - \underline{X}]$ and $(\lozenge(C \cdot Y))_i = [\sum_{j=1}^{n} |C_{ij}| \cdot (\underline{X}_j - \overline{X}_j), \sum_{j=1}^{n} |C_{ij}| \cdot (\overline{X}_j - \underline{X}_j)]$. Therefore the proof is finished if $\sum_{j=1}^{n} |C_{ij}| \cdot (\overline{X}_j - \underline{X}_j) < \overline{X}_i - \underline{X}_i$ for $1 \le i \le n$ holds. Define

$$\overline{y}_j := \begin{cases} \overline{X}_j & \text{if } C_{ij} \ge 0 \\ \underline{X}_j & \text{otherwise} \end{cases} \quad \text{and} \quad \underline{y}_j := \begin{cases} \underline{X}_j & \text{if } C_{ij} \ge 0 \\ \overline{X}_j & \text{otherwise} \end{cases} \tag{8}$$

for some fixed $1 \le i \le n$. Then by assumption

$$\underline{X}_i \le z_i + \sum_{j=1}^{n} C_{ij} \overline{y}_j \le \overline{X}_i \quad \text{and} \quad \underline{X}_i \le z_i + \sum_{j=1}^{n} C_{ij} \underline{y}_j \le \overline{X}_i , \tag{9}$$

where either both left or both right $\le$ can be replaced by <. The left of the second inequality in (9) implies

$$-z_i - \sum_{j=1}^{n} C_{ij} \cdot \underline{y}_j \le -\underline{X}_i \tag{10}$$

and adding the right of the first inequality of (9) and (10) yields

$$\sum_{j=1}^{n} C_{ij} \cdot (\overline{y}_j - \underline{y}_j) < \overline{X}_i - \underline{X}_i \tag{11}$$

because one of both $\leq$ can be replaced by $<$. By definition (8) follows $C_{ij} \cdot (\overline{y}_j - \underline{y}_j) = |C_{ij}| \cdot (\overline{X}_i - \underline{X}_i)$ and therefore demonstrating assertion b) using (11). Assertion a) follows by b).

ad c): Follows by the preceding proof for cases a) and b).

□

Consider a system of linear equations $Ax = b$. For an approximate inverse R of A lemma 3 can be applied to the residual function $f(x) := R \cdot b + \{I - RA\} \cdot x$. If $f(X) \overset{o}{\subseteq} X$ for some $X \in \mathbb{I}V\mathbb{R}$, then by Brouwer's Fixed Point Theorem there is a $\hat{x} \in X$ with $f(\hat{x}) = \hat{x} = Rb + \{I-RA\}\hat{x}$ and therefore $b - A\hat{x} \in \ker R$. If the non-singularity of R could be shown, then the existence of a zero $\hat{x}$ of $Ax - b = 0$ would be demonstrated. In the following we give criteria for the regularity of R and A and show the existence of a fixed point of f and a zero of $Ax - b = 0$ without using Brouwer's Fixed Point Theorem.

<u>Lemma 4</u>. Let $\mathbb{C} \in \mathbb{P}M\mathbb{R}$ and $\emptyset \neq X \in \mathbb{P}V\mathbb{R}$ compact with $0 \in X$. If

$$\mathbb{C} \cdot X \overset{o}{\subseteq} X \tag{12}$$

then $\forall C \in \mathbb{C} : \rho(C) < 1$.

<u>Proof</u>. Let $C \in \mathbb{C}$. For $Y := X + iX$ holds $\mathbb{C} \cdot Y \overset{o}{\subseteq} Y$ and $0 \in \overset{o}{Y}$. Suppose $C \neq (0)$ and let $\lambda \in \mathbb{C}$, $x \in V\mathbb{C}$ be an eigenvalue/eigenvector pair of C. Define

$$T \in \mathbb{P}\mathbb{C} \quad \text{by} \quad T := \{\gamma \in \mathbb{C} \mid \gamma \cdot x \in Y\}.$$

T is not empty because $0 \in Y$ and T is closed and bounded because Y is compact. Therefore there is some $\gamma^* \in \mathbb{C}$ with

$$|\gamma^*| = \max_{\gamma \in T} |\gamma| .$$

Then by (12) we have $\gamma^* \neq 0$ and

$$C \cdot (\gamma^* x) = (\gamma^* \lambda) x \in \overset{o}{Y}.$$

Because $\gamma^* x \in \partial Y$ and the definition of γ^* follows $|\gamma^*| > |\gamma^* \lambda|$ and therefore $|\lambda| < 1$.

□

With these preperations we get the following theorem.

Theorem 5. Let $Z \in \mathbb{P}V\mathbb{R}$, $\mathfrak{C} \in \mathbb{P}M\mathbb{R}$ and $\emptyset \neq X \in \mathbb{P}V\mathbb{R}$ compact. If

$$Z + \mathfrak{C} \cdot X \overset{0}{\subseteq} X \tag{13}$$

then $\forall C \in \mathfrak{C}: \rho(C) < 1$.

Proof. Follows by lemma 2, b) and lemma 4. □

If the inclusion sets are intervals, then weaker conditions suffice to prove a stronger assertion than stated in theorem 5.

Theorem 6. Let $Z \in \mathbb{P}V\mathbb{R}$, $\mathfrak{C} \in \mathbb{P}M\mathbb{R}$ and $X \in \mathbb{I}V\mathbb{R}$. If

$$\lozenge(Z + \mathfrak{C} \cdot X) \subsetneqq X \tag{14}$$

then $\forall C \in \mathfrak{C}: \rho(C) \leq \rho(|C|) < 1$.

Proof. By lemma 3, b) we have $\mathfrak{C} \cdot Y \subseteq \lozenge(\mathfrak{C} \cdot Y) \overset{0}{\subseteq} Y$ for $Y = X \mathbin{\diamondsuit\!\!\!-} X$. If the i-th component of X is $[\underline{X}_i, \overline{X}_i]$, then $Y_i = [\underline{X}_i - \overline{X}_i, \overline{X}_i - \underline{X}_i]$. Therefore $Y_i = -Y_i$ for $1 \leq i \leq n$ and $|\mathfrak{C}| \cdot Y = \mathfrak{C} \cdot Y \overset{0}{\subseteq} Y$ and the assertion follows by lemma 4 and Perron-Frobenius Theory (cf.[45]). □

Given a general set of vectors $x \in \mathbb{P}V\mathbb{R}$, for $Y = X - X$ holds of course $Y = -Y$, but something similar to $Y_i = -Y_i$ is, in general, not true. In theorem 6 with the weaker assumption $\subsetneqq$ instead of $\overset{0}{\subseteq}$ in theorem 5, the stronger assertion holds that the spectral radius of the absolute value of every matrix $C \in \mathfrak{C}$ is less than one. This follows from the special structure of the elements of $\mathbb{I}V\mathbb{R}$. Therefore the fact, that operations in $\mathbb{I}VS$ are easy to implement and are fast, has to be paid by the impossibility to show $\rho(C) < 1$ where $\rho(|C|) \geq 1$ using (14). With spending more computing time for operations in other structures this deficiency can be eliminated. However, in practical cases the matrices C are of the form $I - RA$ where $R \approx A^{-1}$. In our experience never a case occurred where $\rho(I-RA) < 1$ and $\rho(|I-RA|) \geq 1$. On the other hand there are cases where $\|C\| \geq 1$ for the common norms but $X^{k+1} := Z \mathbin{\diamondsuit\!\!\!+} C \mathbin{\diamondsuit\!\!\!\cdot} X^k \overset{0}{\subseteq} X^k$ holds true for $k = 2$ or $k = 3$. Later some examples are shown demonstrating this fact.

With the assumptions of theorem 5 $\rho(|C|) < 1$ for all $C \in \mathfrak{C}$ is, in general, not true. A counterexample is:

$$C := \begin{pmatrix} 1 & -0.25 \\ 3 & -1 \end{pmatrix} \quad \text{with } \rho(C) = 0.5 \text{ and } \rho(|C|) = 1 + \sqrt{0.75} > 1.8 .$$

Define $X := \{\lambda \cdot \begin{pmatrix} 1 \\ 2 \end{pmatrix} + \mu \begin{pmatrix} 1 \\ 6 \end{pmatrix} \mid -1 \le \lambda, \mu \le 1\} \in \mathbb{P}V\mathbb{R}$.

Then $C \cdot X = \{\lambda \cdot \begin{pmatrix} 0.5 \\ 1 \end{pmatrix} + \mu \begin{pmatrix} -0.5 \\ -3 \end{pmatrix} \mid -1 \le \lambda, \mu \le 1\} \overset{\circ}{\subseteq} X$.

The vectors (1,2)' and (1,6)' are the eigenvectors of C with eigenvalues 0.5 and -0.5. It is an open problem whether something similar to $\subsetneqq$ can be defined for $X \in \mathbb{P}V\mathbb{R}$ to weaken the assumptions of Theorem 5. Theorem 5 has been proved for compact and convex sets in [37].

The assumptions of theorem 6 can still be weakened. For preparing this we need the following lemma.

Lemma 7. Let $Z \in \mathbb{P}V\mathbb{R}$, $C \in M\mathbb{R}$ and $X \in \mathbb{I}V\mathbb{R}$. If C is irreducible, $d(X) > 0$ and

$$O(Z + C \cdot X) \tilde{\subseteq} X$$

then $\rho(|C|) < 1$.

Proof. By lemma 3, c), $|C| \cdot Y \subseteq C \diamondminus Y \tilde{\subseteq} Y$ for $Y = X \diamondminus X$. The assertion follows by Perron-Frobenius Theory (cf. [45]). □

Theorem 8. Let $Z \in \mathbb{P}V\mathbb{R}$, $\mathfrak{C} \in \mathbb{P}M\mathbb{R}$ and $X \in \mathbb{I}V\mathbb{R}$. If

$$O(Z + \mathfrak{C} \cdot X) \subsetneqq X \quad \text{using Einzelschrittverfahren,} \tag{15}$$

i.e. for $Y_i := \{O(Z + \mathfrak{C} \cdot (Y_1, \ldots, Y_{i-1}, X_i, \ldots, X_n)^T\}_i$ holds $Y_i \subsetneqq X_i$ for $1 \le i \le n$, then for every $C \in \mathfrak{C}$: $\rho(C) \le \rho(|C|) < 1$.

Proof. Let $z \in Z$ and $C \in \mathfrak{C}$, arbitrarily chosen. By Perron-Frobenius Theory (cf. [45], (2.41) page 46) there is a permutation matrix P with

$$P\,C\,P^T = \begin{pmatrix} R_{11} & R_{12} & \cdots & R_{1n} \\ & R_{22} & \cdots & R_{2n} \\ & & \ddots & \vdots \\ & & & R_{nn} \end{pmatrix}$$

where $R_{ii} \equiv 0$ or R_{ii} irreducible for $1 \leq i \leq n$. Let $1 \leq \nu \leq n$ be fixed but arbitrary. If $R_{\nu\nu} \equiv 0$ then $\rho(R_{\nu\nu}) < 1$; suppose $R_{\nu\nu} \not\equiv 0$ and let $P \cdot (1,2,\ldots,n)^T = (\sigma(1),\sigma(2),\ldots,\sigma(n))^T$, $\tau(\sigma(j)) := j$ for $1 \leq j \leq n$. Let $R_{\nu\nu} \in M_k\mathbb{R}$, $k \leq n$ with indices $\ell,\ldots,\ell+k-1$. The indices of A belonging to $R_{\nu\nu}$ are $\tau(\ell) =: \alpha_1,\ldots,\tau(\ell+k-1) =: \alpha_k, \alpha_m := \max(\alpha_1,\ldots,\alpha_k)$. Because of the irreducibility of $R_{\nu\nu}$ there is a path from α_i to α_m for every $1 \leq i \leq k$ (cf. [45]). Therefore there is a j with $1 \leq j \leq k$ and $C_{\alpha_j,\alpha_m} \neq 0$. For arbitrary $\bar{y}_\mu, \tilde{y}_\mu \in Y_\mu$, $1 \leq \mu \leq i-1$ and $\bar{x}_\mu, \tilde{x}_\mu \in X_\mu$, $i \leq \mu \leq n$ and arbitrary i with $1 \leq i \leq n$, holds

$$(C \cdot (\bar{y}_1-\tilde{y}_1,\ldots,\bar{y}_{i-1}-\tilde{y}_{i-1},\bar{x}_i-\tilde{x}_i,\ldots,\bar{x}_n-\tilde{x}_n)^T)_i =$$

$$(z + C \cdot (\bar{y}_1,\ldots,\bar{y}_{i-1},\bar{x}_i,\ldots,\bar{x}_n)^T)_i - (z + C \cdot (\tilde{y}_1,\ldots,\tilde{y}_{i-1},\tilde{x}_i,\ldots,\tilde{x}_n)^T)_i \subseteq Y_i - Y_i$$

and therefore $(C \cdot (Y_1-Y_1,\ldots,Y_{i-1}-Y_{i-1},X_i-X_i,\ldots,X_n-X_n)^T)_i \subseteq Y_i - Y_i =$

$$= Y_i \diamondminus Y_i \subsetneqq X_i - X_i \,.$$

That means for $U := X - X$

$$C \diamondcdot U \subsetneqq U \quad \text{using Einzelschrittverfahren.} \tag{16}$$

Let $M := \{\alpha_1,\ldots,\alpha_k\}$ and define

$$C^*_{i\ell} := \begin{cases} C_{i\ell} & \text{if } i \in M,\ \ell \in M \\ 0 & \text{otherwise} \end{cases} \quad \text{and} \quad U^*_i := \begin{cases} U_i & \text{if } i \in M \\ 0 & \text{otherwise.} \end{cases}$$

Because $0 \in U_i$, $1 \leq i \leq n$ we have

$$(C^* \cdot U^*)_{\alpha_j} \subseteq (C \cdot U)_{\alpha_j} = (C \cdot (Y_1-Y_1,\ldots,Y_{\alpha_m-1}-Y_{\alpha_m-1},Y_{\alpha_m}-Y_{\alpha_m},\ldots,Y_n-Y_n)^T)_{\alpha_j}$$

$$\overset{\sim}{\subseteq} (C \cdot (Y_1-Y_1,\ldots,Y_{\alpha_j-1}-Y_{\alpha_j-1},X_{\alpha_j}-X_{\alpha_j},\ldots,X_{\alpha_m}-X_{\alpha_m},\ldots,X_n-X_n)^T)_{\alpha_j}$$

because $C_{\alpha_j\alpha_m} \neq 0$ and the definition of α_m. Define $C_M \in M_k\mathbb{R}$ resp. $U_M \in V_k\mathbb{R}$ to be the matrix resp. vector out of C resp. U with indices in M. Then $C_M \cdot U_M \overset{\sim}{\subseteq} U_M$. But C_M is the permuted $R_{\nu\nu}$ so that by lemma 7 the spectral radius if $|R_{\nu\nu}|$ is less than one. Because ν was arbitrary with $1 \leq \nu \leq n$ the theorem is proved. □

Theorem 8 provides a sharp test for $\rho(|C|) < 1$. As compared to the general case of compact inclusion sets $X \in \mathbb{P}V\mathbb{R}$, for $X \in \mathbb{I}V\mathbb{R}$ the weaker inclusion $\subsetneqq$ and the Einzelschrittverfahren modus suffices.

Furthermore, as will be shown later, inclusion using Einzelschritt-verfahren saves computing time and memory. It is an open problem whether something similar can be defined and be achieved for the general case $x \in \mathbb{P}V\mathbb{R}$ compact.

Theorem 9. Let $z \in V\mathbb{R}$, $C \in M\mathbb{R}$ and $X \in \mathbb{I}V\mathbb{R}$, $d(X) > 0$. If

$$\Diamond(z + C\cdot X) \supseteq X \tag{17}$$

then $\rho(|C|) \geq 1$.

Proof. Let m(X) be the midpoint of X and $Y := X - m(X)$. Then $Y \in \mathbb{I}V\mathbb{R}$, $0 \in Y$ and

$$\Diamond(z + C\cdot X) = z + C \diamondsuit X = z + C \diamondsuit (m(X)+Y) = z + C\cdot m(X) + C \diamondsuit Y \supseteq X \quad \text{and}$$

$$C \diamondsuit Y \supseteq Y + m(X) - z - C\cdot m(X).$$

Because $Y \in \mathbb{I}V\mathbb{R}$ and $Y = -Y$ this implies (abbreviating $v := m(X)-z-C\cdot m(X)$)

$$-|C|\cdot|Y| \leq -|Y| + v \leq |Y| + v \leq |C|\cdot|Y|$$

and therefore

$$|C|\cdot|Y| \geq |Y| \,. \tag{18}$$

Observing $|Y| > 0$ the proof is finished by applying exercise 2, p.47 in [45]. □

If an iteration $X^{k+1} := \Diamond(z + \mathfrak{C}\cdot X^k)$ is performed, theorem 9 gives a stopping criterion because $X^{k+1} \overset{\circ}{\subseteq} X^k$ would imply $\rho(|C|) < 1$ for every $C \in \mathfrak{C}$. Theorem 9 holds true because of the special shape of interval vectors. However, (17) does not imply $\rho(C) > 1$ because of interval dependencies. A counterexample is:

$$C := \begin{pmatrix} 1 & -1 \\ 1 & -1 \end{pmatrix} \quad \text{for } z = 0 \text{ and } X := \left\{\begin{pmatrix} x_1 \\ x_2 \end{pmatrix} \,\middle|\, -1 \leq x_1, x_2 \leq 1\right\}.$$

Then $z + C\cdot X = \left\{\begin{pmatrix} x_1 \\ x_2 \end{pmatrix} \,\middle|\, -2 \leq x_1 = x_2 \leq 2\right\}$, but $z + C \diamondsuit X = \left\{\begin{pmatrix} x_1 \\ x_2 \end{pmatrix} \,\middle|\, -2 \leq x_1, x_2 \leq 2\right\} \supseteq X$ with $\rho(C) = 0$, where $\rho(|C|) = 2$.

Applying iteration schemes to $f(x) = Ax - b$ the aim is to demonstrate the existence (and maybe uniqueness) of a zero of f(x) within a certain domain. A preparation for this is the following lemma.

Lemma 10. Let $z \in V\mathbb{R}$, $C \in M\mathbb{R}$ and $\emptyset \neq X \in \mathbb{P}V\mathbb{R}$ compact. If $\rho(C) < 1$ and

$$z + C \cdot X \subseteq X \tag{19}$$

then there is one and only one $\hat{x} \in V\mathbb{R}$ with $z + C\cdot\hat{x} = \hat{x}$. It is $\hat{x} = (I-C)^{-1}\cdot z \in X$.

Proof. Define f: $V\mathbb{R} \in V\mathbb{R}$ by $f(x) := z + C\cdot x$, then for $m \in \mathbb{N}$

$$f^m(x) = \sum_{i=0}^{m-1} C^i z + C^m x \quad \text{and} \quad (I-C)\cdot f^m(x) = z + C^m\cdot(x-z-Cx).$$

By (19) $f^m(x) \in X$ for $x \in X$ and therefore $(I-C)\cdot f^m(x) \to z$ for every $x \in X$, $m \to \infty$ because C is convergent. But I-C is invertible and X closed and bounded, therefore $f^m(x) \to (I-C)^{-1}\cdot z$ for $m \to \infty$, i.e. $\hat{x} := (I-C)^{-1}\cdot z \in X$. Now

$$f(\hat{x}) - \hat{x} = z + C\cdot(I-C)^{-1}\cdot z - (I-C)^{-1}\cdot z = 0 \tag{20}$$

Suppose $z + C\cdot\hat{y} = \hat{y}$. Then

$$C\cdot(\hat{x} - \hat{y}) = (z + C\cdot\hat{x}) - (z + C\cdot\hat{y}) = \hat{x} - \hat{y} \tag{21}$$

implying $\hat{x} - \hat{y} = 0$ because $\rho(C) < 1$. □

With the previons theorems 5 and 8 we have the tool to verify $\rho(C) < 1$ by means of conditions (13) and (15), which can be implemented on computers. This leads to the following.

Theorem 11. Let $Z \in \mathbb{P}V\mathbb{R}$, $\mathfrak{C} \in \mathbb{P}M\mathbb{R}$ and $\emptyset \neq X \in \mathbb{P}V\mathbb{R}$ compact. If

$$Z + \mathfrak{C}\cdot X \overset{\circ}{\subseteq} X \tag{22}$$

then for every $z \in Z$ and for every $C \in \mathfrak{C}$ holds $\rho(C) < 1$ and there is one and only one $\hat{x} \in V\mathbb{R}$ with $z + C\cdot\hat{x} = \hat{x}$. It is $\hat{x} = (I-C)^{-1}\cdot z \in \mathring{X}$.

Proof. Let $z \in Z$ and $C \in \mathfrak{C}$ fixed but arbitrary. Then (22) and theorem 5 yields $\rho(C) < 1$ and (19) and by lemma 10 the existence of a $\hat{x} \in X$ with $z + C\cdot\hat{x} = \hat{x}$. Moreover $\hat{x} = z + C\cdot\hat{x} \in Z + \mathfrak{C}\cdot X \subseteq \mathring{X}$ by (22). □

Theorem 12. Let $Z \in \mathbb{P}V\mathbb{R}$, $\mathfrak{C} \in \mathbb{P}M\mathbb{R}$ and $X \in \mathbb{I}M\mathbb{R}$. If

$$\lozenge(Z + \mathfrak{C}\cdot X) \subsetneqq X \quad \text{using Einzelschrittverfahren,} \tag{23}$$

i.e. for $Y_i := \{\lozenge(Z + \mathbb{C}\cdot(Y_1,\dots,Y_{i-1},X_i,\dots,X_n)^T\}_i$ holds $Y_i \subsetneqq X_i$ for $1 \le i \le n$, then for every $z \in Z$ and for every $C \in \mathbb{C}$ holds $\rho(|C|) < 1$ and there is one and only one fixed point $\hat{x} \in V\mathbb{R}$ with $z + C\cdot\hat{x} = \hat{x}$. It is $\hat{x} = (I-C)^{-1}\cdot z \in Y$ and $Z + \mathbb{C}\cdot Y \subseteq Y$.

Proof. Let $z \in Z$ and $C \in \mathbb{C}$ fixed but arbitrary. Then (23) and theorem 8 implies $\rho(C) < 1$. Moreover

$$\{z + C\cdot(Y_1,\dots,Y_n)^T\}_i \subseteq \{z + C\cdot(Y_1,\dots,Y_{i-1},X_i,\dots,X_n)^T\}_i = Y_i \text{ for } 1 \le i \le n$$

implying $z + C\cdot Y \subseteq Y$. Lemma 10 concludes the proof. □

The next theorem gives another improvement of the inclusion assumption (13) in theorem 5, combined with an iteration scheme.

Theorem 13. Let $Z \in \mathbb{P}V\mathbb{R}$, $\mathbb{C} \in \mathbb{P}M\mathbb{R}$ and $\emptyset \neq X \in \mathbb{P}V\mathbb{R}$, all Z, $\mathbb{C}$ and X being compact. Define $f: \mathbb{P}V\mathbb{R} \to \mathbb{P}V\mathbb{R}$ by $f(V) := Z + \mathbb{C}\cdot V$ for $V \in \mathbb{P}V\mathbb{R}$. If

$$f^{k+m}(X) \overset{\circ}{\subseteq} f^k(X) \quad \text{for some } k,\ m \in \mathbb{N},\ k \ge 0,\ m \ge 1 \tag{24}$$

then for every $z \in Z$ and for every $C \in \mathbb{C}$ holds $\rho(C) < 1$ and there is one and only one $\hat{x} \in V\mathbb{R}$ with $z + C\cdot\hat{x} = \hat{x}$. It is $\hat{x} = (I-C)^{-1}\cdot z \overset{\circ}{\in} f^k(X)$.

Proof. Let $z \in Z$ and $C \in \mathbb{C}$ fixed but arbitrary. Define $g: V\mathbb{R} \to V\mathbb{R}$ by $g(x) := z + C\cdot x$ and let $Y := f^k(X)$. Y is compact because Z, $\mathbb{C}$ and X are, so with

$$g^m(Y) = \sum_{i=0}^{m-1} C^i\cdot z + C^m\cdot Y \subseteq f^m(Y) \overset{\circ}{\subseteq} Y, \tag{25}$$

theorem 5 implies $\rho(C^m) < 1$ and therefore $\rho(C) < 1$, and $g^m(\hat{x}) = \hat{x}$ for $\hat{x} := (I-C^m)^{-1}\cdot(\sum_{i=0}^{m-1} C^i)\cdot z$. On the other hand

$$(\sum_{i=0}^{m-1} C^i)(I-C) = I - C^m, \tag{26}$$

and by $\rho(C) < 1$ and $\rho(C^m) < 1$ follows $\hat{x} = (I-C)^{-1}\cdot z \in Y$. Now (20) and (21) complete the proof. □

When aiming to compute an inclusion of the solution of a given problem (e.g. a system of linear of nonlinear equations), an iteration scheme is very useful. Since the problem of computing an inclusion is later reduced to verify contraction for an affine mapping, let us consider (24) more closely. Consider the one-dimensional, contracting affine function

$$f(x) := 3 - 0.5 \cdot x. \tag{27}$$

For $X := [1.9, 2.1]$ is $f(X) = [1.95,\ 2.05] \overset{\circ}{\subseteq} X$. Therefore f has one and only one fixed point by theorem 11 and $\hat{x} = (1-(-0.5))^{-1} \cdot 3 = 2$. By the formula used in (25) is

$$f^m(x) = \sum_{i=0}^{m-1} (-0.5)^i \cdot 3 + (-0.5)^m \cdot x = \frac{1-(-0.5)^m}{1-(-0.5)} \cdot 3 + (-0.5)^m \cdot x = 2 + (-0.5)^m \cdot (x-2).$$

For $X^0 := [1.9, 2.3]$ and $X^m := f^m(X)$ is

$$X^1 = [1.85, 2.05],\ X^2 = [1.975, 2.075],\ X^3 = [1.9625, 2.0125], \ldots$$

It is easy to show $X^{m+1} \not\subseteq X^m$ for every $0 \le m \in \mathbb{N}$. Therefore the contraction cannot be established by theorem 11. However, $X^2 \overset{\circ}{\subseteq} X^0$ and theorem 13 can be applied. For $X^0 := [3,5]$ is

$$X^1 = [0.5, 1.5],\ X^2 = [2.25, 2.75] \text{ and } X^m = 2 + (-0.5)^m \cdot [1,3]. \tag{28}$$

Therefore $2 \notin X^m$ for every $0 \le m \in \mathbb{N}$ and by theorem 13, (24) can never be satisfied. Therefore, although the function f defined by (27) is contracting, no inclusion is possible using any of the theorems so far (for the Einzelschrittverfahren as well since the function is one-dimensional).

Next we seek for methods providing inclusions even under extreme circumstances. The next theorems allows an inclusion for the example above.

Theorem 14. Let $Z \in \mathbb{P}V\mathbb{R}$, $\mathfrak{C} \in \mathbb{P}M\mathbb{R}$ and $\emptyset \neq X \in \mathbb{P}V\mathbb{R}$, all $Z, \mathfrak{C}$ and X being compact. Define $f : \mathbb{P}V\mathbb{R} \to \mathbb{P}V\mathbb{R}$ by $f(V) := Z + \mathfrak{C} \cdot V$ for $V \in \mathbb{P}V\mathbb{R}$. If

$$f^{m+1}(X) \overset{\circ}{\subseteq} \bigcup_{i=0}^{m} f^i(X) \quad \text{for some} \quad 0 \le m \in \mathbb{N}, \tag{29}$$

then for every $z \in Z$ and for every $C \in \mathfrak{C}$ holds $\rho(C) < 1$ and there is one and only one $\hat{x} \in V\mathbb{R}$ with $z + C \cdot \hat{x} = \hat{x}$. It is $\hat{x} = (I-C)^{-1} \cdot z \in \bigcup_{i=0}^{m} f^i(X)$.

Proof. For $U, V \in \mathbb{P}V\mathbb{R}$ is

$$f(U \cup V) = f(U) \cup f(V).$$

For $Y := \bigcup_{i=0}^{m} f^i(X)$ is $f^{m+1}(X) \overset{\circ}{\subseteq} Y$ and by induction for $k \ge 1$

$$f^{m+k+1}(X) \subseteq f(f^{m+k}(X)) \overset{o}{\subseteq} f(Y) \subseteq f^{m+1}(X) \cup \underline{\bigcup}_{i=0}^{m} f^{i}(X) \subseteq Y. \tag{30}$$

Therefore

$$f^{m+1}(Y) = \underline{\bigcup}_{i=0}^{m} f^{m+1}(f^{i}(X)) = \underline{\bigcup}_{i=0}^{m} f^{m+i+1}(X) \overset{o}{\subseteq} Y.$$

Y is compact because Z, $\mathfrak{C}$ and X are and applying theorem 13 completes the proof. □

<u>Corollary 15</u>. Let $Z \in \mathbb{P}V\mathbb{R}$, $\mathfrak{C} \in \mathbb{P}M\mathbb{R}$ and $\emptyset \neq X \in \mathbb{P}V\mathbb{R}$, all Z, $\mathfrak{C}$ and X being compact. Define $f: \mathbb{P}V\mathbb{R} \to \mathbb{P}V\mathbb{R}$ by $f(V) := Z + \mathfrak{C} \cdot V$ for $V \in \mathbb{P}V\mathbb{R}$. If ($\underline{\cup}$ denotes the convex union)

$$f^{m+1}(X) \overset{o}{\subseteq} \underline{\bigcup}_{i=0}^{m} f^{i}(X) \quad \text{for some} \quad 0 \leq m \in \mathbb{N},$$

then for every $z \in Z$ and every $C \in \mathfrak{C}$ holds $\varrho(C) < 1$ and there is one and only one $\hat{x} \in V\mathbb{R}$ with $z + C \cdot \hat{x} = \hat{x}$. It is $\hat{x} = (I-C)^{-1} \cdot z \in \underline{\bigcup}_{i=0}^{m} f^{i}(X)$.

<u>Proof</u>. Follows by replacing $\cup$ by $\underline{\cup}$ in the proof of theorem 14 or, by lemma 2,a) and theorem 13. □

Note, that corollary 15 holds true because f is an affine function. Let $A, B \in \mathbb{P}V\mathbb{R}$, then

$$\begin{aligned} f(A \underline{\cup} B) &= \{f(x) \mid x = a + \lambda(b-a) \text{ for } a \in A,\ b \in B,\ \lambda \in \mathbb{R} \text{ with } 0 \leq \lambda \leq 1\} = \\ &= \{z + C \cdot (a + \lambda(b-a)) \mid a \in A,\ b \in B,\ \lambda \in \mathbb{R} \text{ with } 0 \leq \lambda \leq 1\} = \\ &= \{z + C \cdot a + \lambda \cdot ((z + C \cdot b) - (z + C \cdot a)) \mid a \in A,\ b \in B,\ \lambda \in \mathbb{R} \text{ with } 0 \leq \lambda \leq 1\} = \\ &= \{z + C \cdot a \mid a \in A\} \underline{\cup} \{z + C \cdot b \mid b \in B\} = \\ &= f(A) \underline{\cup} f(B). \end{aligned}$$

If the function f is not affine, there need not be a relation between $f(A \underline{\cup} B)$ and $f(A) \underline{\cup} f(B)$. Consider as examples:

1.) $f: \mathbb{R} \to \mathbb{R}$ with $f(x) = x$ and $A := \{1\}$, $B = \{-1\}$. Then

$f(A \underline{\cup} B) = \{f(x) \mid x \in [-1,1]\} = [0,1]$ and

$f(A) \underline{\cup} f(B) = \{1\} \underline{\cup} \{1\}$ with $f(A \underline{\cup} B) \not\subseteq f(A) \underline{\cup} f(B)$.

2.) $f:\mathbb{R}^2 \to \mathbb{R}^2$ with $f(x,y) = (x^2,y^3)^T$ and $A = \{\binom{0}{0}\}$, $B = \{\binom{1}{1}\}$. Then

$f(A \underline{\cup} B) = \{f(x,y) \mid 0 \le x = y \le 1\} = \{\binom{x^2}{x^3} \mid 0 \le x \le 1\}$ and

$f(A) \underline{\cup} f(B) = \{\binom{0}{0}\} \underline{\cup} \{\binom{1}{1}\} = \{\binom{x}{x} \mid 0 \le x \le 1\}$ with

$f(A) \underline{\cup} f(B) \not\subseteq f(A \underline{\cup} B)$ and $f(A \underline{\cup} B) \not\subseteq f(A) \underline{\cup} f(B)$. In fact

$(f(A) \underline{\cup} f(B)) \cap f(A \underline{\cup} B) = \{\binom{0}{0},\binom{1}{1}\} = f(A) \cup f(B)$ which is the bare minimum the intersection must contain.

For the function f defined in (27) and $X := [3,5]$ is by (28)

$$f^2(X) = [2.25,2.75] \overset{\circ}{\subseteq} X \underline{\cup} f(X) = [3,5] \underline{\cup} [0.5,1.5] = [0.5,5].$$

This implies f to be contractive and gives a (poor) inclusion $[0.5,5]$ for the fixed point $\hat{x} = 2$ of f. In the following chapter theorems are given for improving the quality of an inclusion.

3. Linear systems. Our next aim are theorems for inclusions of the solution of systems of linear equations of those theorems should be verifiable on computers such that inclusions sets can be calculated. In [17] and [39] a number of those theorems are given. Next those theorems will be improved. The inclusion formula of the next theorem occured in [21]. There, Krawczyk supposed $\rho(I-RA) < 1$ a priori. Using the proper inclusion $\overset{\circ}{\subseteq}$ we have an easy verifaction scheme for the iteration matrix to be convergent.

Theorem 16. Let $A,R \in M\mathbb{R}$, $b,\tilde{x} \in V\mathbb{R}$. If for some compact $\emptyset \neq X \in \mathbb{P}V\mathbb{R}$

$$\tilde{x} + R(b - A\tilde{x}) + \{I - RA\}\cdot(X-\tilde{x}) \overset{\circ}{\subseteq} X, \tag{31}$$

then the matrices A and R are not singular and the uniquely determined solution $\hat{x} := A^{-1}\cdot b$ of $Ax = b$ satisfies $\hat{x} \in \overset{\circ}{X}$.

Proof. Short computation yields for $x \in V\mathbb{R}$

$$\tilde{x} + R(b-A\tilde{x}) + \{I-RA\}\cdot(x-\tilde{x}) = R\cdot b + \{I - RA\}\cdot x \,. \tag{32}$$

Therefore $R\cdot b + \{I-RA\}\cdot X \overset{\circ}{\subseteq} X$ and by theorem 11 $\rho(I-RA) < 1$ proving the non-singularity of R and A. The fixed point $\hat{x}$ of the function $R\cdot b + \{I-RA\}\cdot x$ satisfies $\hat{x} \in \overset{\circ}{X}$ and $\hat{x} = \{I-(I-RA)\}^{-1}\cdot R\cdot b = A^{-1}\cdot b$ by theorem 11.

□

Obviously (31) can be replaced by $R\cdot b + \{I-RA\}\cdot X \overset{\circ}{\subseteq} X$, which is, for appropriate X, also verifiable on computers.

One significant improvement of the quality of inclusions is not to include the solution itself but the difference to an approximate solution (cf. [37] and [38]). As shown by (32) the approximation $\tilde{x}$ to $\hat{x}$ does not play any role in (31). If, instead the inclusion shall satisfy $\hat{x} - \tilde{x} \in X$, then the linear system $Ax = b-A\tilde{x}$ has to be solved because

$$A(\hat{x} - \tilde{x}) = A\hat{x} - A\tilde{x} = b - A\tilde{x} \,.$$

This leads to the well-known residue iteration scheme. The corresponding inclusion theorem is the following.

Theorem 17. Let $A,R \in M\mathbb{R}$, $b,\tilde{x} \in V\mathbb{R}$. If for some compact $\emptyset \neq X \in \mathbb{P}V\mathbb{R}$

$$R\cdot(b-A\tilde{x}) + \{I-RA\} \cdot X \overset{\circ}{\subseteq} X, \tag{33}$$

then the matrices A and R are not singular and the uniquely determined solution $\hat{x} := A^{-1}\cdot b$ of $Ax = b$ satisfies $\hat{x} \in \tilde{x} + X$. If

$$R\cdot(b-A\tilde{x}) + \{I-RA\}\cdot X \cap X = \emptyset \,, \tag{34}$$

then there is no solution of $Ax = b$ in $\tilde{x} + X$.

Proof. The first assertion follows by applying theorem 16 replacing b by $b-A\tilde{x}$ and regarding (32). Suppose $A\hat{x} = b-A\tilde{x}$. Then

$$R\cdot(b-A\tilde{x}) + \{I-RA\}\cdot(\hat{x}-\tilde{x}) = \hat{x}-\tilde{x} \tag{35}$$

would contradict (34) if $\hat{x} - \tilde{x} \in X$.

□

Theorem 17 can be applied on computers if operations extending the power set operations, possibly for special inclusion sets, are available. One possibility is to restrict the inclusion sets to interval vectors, i.e. hyperrectangles or, in the complex number space, to n-dimensional balls or torus-sectors. For the application on computers we formulate theorem 17 for interval vectors over some subset G of $\mathbb{R}$. $\lozenge : \mathbb{P}V\mathbb{R} \to \mathbb{I}VG$ is some isotone rounding, i.e. satisfying $X \in \mathbb{P}V\mathbb{R} \Rightarrow X \subseteq \lozenge X$.

Theorem 18. Let $G \subseteq \mathbb{R}$ be some subset of $\mathbb{R}$, $A,R \in MG$, $b,\tilde{x} \in VG$ and $X \in \mathbb{I}VG$. If

$$\lozenge(R\cdot(b-A\tilde{x}) + \{I-RA\}\cdot X) \subsetneq X \quad \text{using Einzelschrittverfahren,} \tag{36}$$

i.e. for $Y_i := \{\lozenge(R(b-A\tilde{x}) + \{I-RA\}\cdot(Y_1,\ldots,Y_{i-1}, X_i,\ldots,X_n)^T\}_i$ holds $Y_i \subsetneq X_i$ for $1 \le i \le n$, then the matrices A and R are not singular and the uniquely determined solution $\hat{x} := A^{-1}\cdot b$ of $Ax = b$ satisfies $\hat{x} \in \tilde{x} + Y$.

Proof. By theorem 12 is $\rho(I-RA) < 1$ and therefore A and R regular, and $\{I-(I-RA)\}^{-1}\cdot R\cdot(b-A\tilde{x}) = A^{-1}\cdot(b-A\tilde{x}) = A^{-1}\cdot b - \tilde{x} \in Y$. □

Assumption (36) can be verified on computers. If G is a set of floating-point numbers and operations $\circledast : \mathbb{I}G \to \mathbb{I}G$ satisfying $A,B \in \mathbb{I}G$: $A * B \subseteq A \circledast B$ for $* \in \{+,-,\cdot,/\}$ are available, then

$$R \odot (b \ominus A \odot \tilde{x}) \oplus \{I \ominus R \odot A\} \odot X \subsetneq X \quad \text{using Einzelschrittverfahren} \tag{37}$$

implies (36) and therefore the assertions of theorem 18 hold true. If a precise scalar product is available, then

$$R \odot \lozenge\{b-Ax\} \oplus \lozenge\{I-RA\} \odot X \subsetneq X \quad \text{using Einzelschrittverfahren} \tag{38}$$

can be used. Obviously, (38) implies (36) but is much sharper then (37) because the critical parts, the residuals, are enclosed with maximum accuracy.

It occurs frequently, that either the input data of a numerical problem is not exactly representable in a given floating-point screen or, that input data is afficted with tolerances. In the latter case, one is interested in an inclusion of the set of all possible solutions for all possible combinations of input data. In the first case, the input data could be enclosed in the input range of the immediate predecessor and successor in the floating-point grid and the resulting problem with input affected with tolerances be solved. For input data with tolerances consider the following theorem.

Theorem 19. Let $\mathbf{A} \in \mathbb{P}M\mathbb{R}$, $\mathbf{b} \in \mathbb{P}V\mathbb{R}$, $R \in M\mathbb{R}$ and $\tilde{x} \in V\mathbb{R}$. If for some compact $\emptyset \neq X \in \mathbb{P}V\mathbb{R}$

$$R\cdot(\mathbf{b}-\mathbf{A}\tilde{x}) + \{I-R\mathbf{A}\}\cdot X \overset{\circ}{\subseteq} X, \tag{39}$$

then for every $A \in \mathbf{A}$ and for every $b \in \mathbf{b}$ the following is true: the matrices A and R are not singular and the uniquely determined solution $\hat{x} := A^{-1}b$ of $Ax = b$ satisfies $\hat{x} \in \tilde{x} + X$.

Proof. Follows by applying theorem 17 to every $A \in \mathbf{A}$, $b \in \mathbf{b}$. □

For immediate applications on computers the power set operations can be replaced by corresponding isotone operations, which are executable on computers. The input data $\mathbf{A}$ and $\mathbf{b}$, which are affected with tolerances, can be replaced by computer representable sets. Usually, one would choose R and $\tilde{x}$ to be point data. An example of such a theorem is the following.

Theorem 20. Let $G \subseteq \mathbb{R}$ be some subset of $\mathbb{R}$, $\mathbf{A} \in \mathbb{I}MG$, $\mathbf{b} \in \mathbb{I}VG$, $R \in MG$ and $\tilde{x} \in VG$. If for some $X \in \mathbb{I}VG$

$$R \odot \lozenge(\mathbf{b}-\mathbf{A}\tilde{x}) \oplus \lozenge\{I-R\mathbf{A}\} \odot X \subsetneqq X \text{ using Einzelschrittverfahren,}$$

then for every $A \in M\mathbb{R}$ and $b \in V\mathbb{R}$ with $A \in \mathbf{A}$ and $b \in \mathbf{b}$ the following is true: the matrices A and R are not singular and the uniquely determined solution $\hat{x} := A^{-1}b$ of $Ax = b$ satisfies $\hat{x} \in \tilde{x} \oplus X$.

Proof. Follows by applying theorem 18 to every $A \in \mathbf{A}$, $b \in \mathbf{b}$. □

Note, that the assertion of theorem 20 applies to every real matrix $A \in M\mathbb{R}$ and every real vector $b \in V\mathbb{R}$ with $A \in \mathbf{A}$, $b \in \mathbf{b}$ and not only to those with elements in G. Moreover, a computer verification holds true without any restriction, e.g. with respect to a given floating-point precision.

In [38] an algorithm is given for computing an inclusion of the solution of a system of linear equations. After computing an approximate inverse R of A a residual correction method is applied yielding some approximate solution $\tilde{x}$. After computing an inclusion for $\mathfrak{C} := \lozenge(I-RA)$ an iteration is started. To illustrate the differences between the different iteration schemes for obtaining an inclusion consider the following example. Let

$$A := \begin{pmatrix} 1 & 2 \\ 2 & 3 \end{pmatrix} ; \quad A^{-1} = \begin{pmatrix} -3 & 2 \\ 2 & -1 \end{pmatrix} ; \quad R := \begin{pmatrix} -2.85 & 2.1 \\ 1.85 & -1.1 \end{pmatrix} . \text{ Then}$$

$$C := \begin{pmatrix} -0.35 & -0.6 \\ 0.35 & 0.6 \end{pmatrix} \quad \text{and} \quad \rho(|C|) = 0.95 ; \quad b := \begin{pmatrix} 1 \\ 0 \end{pmatrix} . \tag{40}$$

The following computational results are obtained on a IBM S/370 processor in single precision, i.e. 6 hexadecimal digits or approximately 7 decimal digits. We apply three different iteration schemes according to theorems 16,17 and 18:

$$\begin{aligned} &1)\ Z := \tilde{x} \oplus R \odot \lozenge\{b-A\tilde{x}\}; \quad X^o := Z; \quad X^{k+1} := Z \oplus C \odot (X^k \ominus \tilde{x}); \\ &2)\ Z := R \odot b; \quad X^o := Z; \quad X^{k+1} := Z \oplus C \odot X^k; \\ &3)\ Z := R \odot \lozenge\{b-A\tilde{x}\}; \quad X^o := Z; \quad X^{k+1} := Z \oplus C \odot X^k; \end{aligned} \tag{41}$$

Then $X^{k+1} \subsetneqq X^k$ proves the non-singularity of A and R and $\hat{x} \in X$ resp. $\hat{x} \in x \oplus X$ in cases 1) and 2) resp. 3), $\hat{x} := A^{-1} \cdot b$ according to theorems 16,17 and 18.

In the following table from left to right the number i of the iteration scheme, the number of iterations k and the maximum relative error e for the two inclusions (defined by $\max\limits_{x_1,x_2 \in X} \left|\frac{x_1-x_2}{x_1}\right|$ for an interval X) are displayed.

i	k	e
1	no inclusion	
2	10	$1.5 \cdot 10^{-5}$
3	1	$6.0 \cdot 10^{-8}$

For the first iteration, no inclusion is obtained because the iterates X^k become cyclic. The third scheme enclosing the difference $\hat{x} - \tilde{x}$ is superior with respect to the number of iterations necessary and with respect to the quality of the inclusion. Nevertheless it is possible that $\rho(|C|) < 1$ but neither of the iteration schemes (41) obtain an inclusion. Consider

$$f(x) := 3 + 0.5 \cdot x. \tag{42}$$

Then by the formula used in (25)

$$f^m(x) = \sum_{i=0}^{m-1} 0.5^i \cdot 3 + 0.5^m \cdot x = \frac{1-0.5^m}{1-0.5} \cdot 3 + 0.5^m \cdot x = 6 + 0.5^m \cdot (x-6).$$

For $X^o := [4,5]$ is

$$X^1 = [5,5.5]; \; X^2 = [5.5,5.75]; \; X^3 = [5.75,5.875] \text{ etc.}$$

It is easy to see that $X^{k+m} \not\subseteq X^k$ and $X^{k+m} \not\subseteq \bigcup_{i=0}^{m-1} X^{k+i}$ for every $k,m \in \mathbb{N}$.

For the purpose of obtaining an inclusion even in those cases, in [37] the ε-inflation has been introduced. An iteration with ε-inflation is

$$\underline{\text{repeat}} \; Y := X \circ \varepsilon; \; X := Z + C \cdot Y$$
$$\underline{\text{until}} \quad X \overset{\circ}{\subseteq} Y \tag{43}$$

for some starting interval X and the property $X \in \mathbb{P}V\mathbb{R} : X \subseteq X \circ \varepsilon$. One could, e.g., define $X \circ \varepsilon := X + [-0.5,+0.5]$. Then with $X := [4,5]$ for the example in (42)

$$Y = [3.5,5.5]; \; X = [4.75,5.75]; \; Y = [4.25,6.25]; \; X = [5.125,6.125]$$

with an inclusion [5.125,6.125] and all the verifications. When including the difference to an approximate solution rather than the solution itself this inclusion will be sharpened.

What is observed in the example can be generalized. It can be shown, that an iteration (43) terminates if and only if $\rho(C) < 1$. It is clear from the previous theorems, that if (43) terminates then $\rho(C) < 1$. The other direction is demonstrated in the following.

Lemma 21. Let $Z \in \mathbb{P}V\mathbb{R}$, $C \in M\mathbb{R}$ and $E_i \in \mathbb{P}V\mathbb{R}$, $i \geq 1$. Suppose Z and the E_i to be bounded, $E_{i+1} \subseteq E_i$ and $U_\varepsilon(0) \subseteq E_i$ for all $i \geq 1$ and some fixed $0 < \varepsilon \in \mathbb{R}$. Define $f: \mathbb{P}V\mathbb{R} \to \mathbb{P}V\mathbb{R}$ by

$$X \in \mathbb{P}V\mathbb{R} : \quad f(X) := Z + C \cdot X. \tag{44}$$

For some $X^o \in \mathbb{P}V\mathbb{R}$ bounded define

$$X^{k+1} := f(X^k) + E_{k+1} \quad \text{for} \quad 0 \leq k \in \mathbb{N}. \tag{45}$$

If $\rho(C) < 1$ then there is some $m \in \mathbb{N}$ with

$$f(X^m) \overset{o}{\subseteq} X^m . \tag{46}$$

Proof. First we proof be induction

$$X^m = \sum_{i=0}^{m-1} C^i \cdot (Z + E_{m-i}) + C^m \cdot X^o \quad \text{for} \quad 0 \leq m \in \mathbb{N}. \tag{47}$$

(47) is true for $m = 0$. Supposing (47) to hold for $m \in \mathbb{N}$ yields by definition (44) and (45)

$$X^{m+1} = Z + E_{m+1} + C \cdot \sum_{i=0}^{m-1} C^i (Z+E_{m-i}) + C^{m+1} \cdot X^o = \sum_{i=0}^{m} C^i (Z + E_{m+1-i}) + C^{m+1} \cdot X^o. \tag{48}$$

Because C is contracting there is an $m \in \mathbb{N}$ satisfying

$$C^m \cdot (Z+E_1) + C^{m+1} \cdot X^o - C^m \cdot X^o \overset{o}{\subseteq} E_{m+1}$$

because $U_\varepsilon(0) \subset E_i$, $i \geq 1$. This implies

$$Z + \sum_{i=1}^{m-1} C^i \cdot (Z+E_{m+1-i}) + C^m (Z+E_1) + C^{m+1} \cdot X^o \overset{o}{\subseteq} Z + E_{m+1} + \sum_{i=1}^{m-1} C^i \cdot (Z+E_{m+1-i}) + C^m \cdot X^o$$

(for $A,B,C \in \mathbb{P}V\mathbb{R}$: $A - B \subseteq C$ implies $A \subseteq B + C$, but the contrary is not true).
Therefore

$$Z + C \cdot \{ \sum_{i=0}^{m-1} C^i \cdot (Z+E_{m-i}) + C^m \cdot X^o \} \overset{o}{\subseteq} \sum_{i=0}^{m-1} C^i \cdot (Z+E_{m+1-i}) + C^m \cdot X^o.$$

Regarding (47) and $E_{i+1} \subseteq E_i$ proofs the lemma. □

Theorem 22. Let $Z \in \mathbb{P}V\mathbb{R}$, $C \in M\mathbb{R}$ and $E_i \in \mathbb{P}V\mathbb{R}$ for $i \geq 1$, all Z and E_i being compact. Suppose $E_{i+1} \subseteq E_i$ and $U_\varepsilon(0) \subseteq E_i$ for all $i \geq 1$ and some fixed $0 < \varepsilon \in \mathbb{R}$. Define $f : \mathbb{P}V\mathbb{R} \to \mathbb{P}V\mathbb{R}$ by

$$X \in \mathbb{P}V\mathbb{R}: \quad f(X) := Z + C \cdot X.$$

For some compact $X^o \in \mathbb{P}V\mathbb{R}$ define

$$X^{k+1} := f(X^k) + E_{k+1} \quad \text{for} \quad 0 \leq k \in \mathbb{N}.$$

Then the following is equivalent:

(1) $f(X^m) \overset{o}{\subseteq} X^m$ for some $m \in \mathbb{N}$

(2) $\rho(C) < 1$.

Proof. (1) ⇒ (2) follows by theorem 5 and the fact, that X^m is compact and nonempty. (2) ⇒ (1) follows by lemma 21. □

Theorem 22 shows that whenever in inclusion is possible, namely if the iteration matrix is convergent, an inclusion will be achieved using an iteration scheme (43) with the ε-inflation. This equivalence is a best possible result, it is the equivalent to the same condition for a real iteration

$$x^{k+1} := x^k + R\cdot(b - Ax^k)$$

for the linear system $Ax = b$, $R \approx A^{-1}$. Similar conditions for $\rho(|C|) < 1$ in case $X^o \in \mathbb{II}V\mathbb{R}$ are given in [39].

In practice, the ε-inflation would be dependent on the iterate. For more details see [38], [39].

Theorem 22 does not necessarily hold true for sets of matrices. In fact it may happen that for a convex set of matrices $\mathfrak{C}$ every $C \in \mathfrak{C}$ is convergent but $C_1 \cdot C_2$ for $C_1, C_2 \in \mathfrak{C}$ is not. An example is

$$\mathfrak{C} := \{A + \sigma(B-A) \mid 0 \le \sigma \le 1\} \text{ for } A := \begin{pmatrix} 0.5 & 0.27 \\ 0.92 & 0.5 \end{pmatrix}, \quad B := \begin{pmatrix} 0.17 & 0.6 \\ 0.94 & 0.25 \end{pmatrix}.$$

Then $\rho(A) \approx 0.9984$, $\rho(B) \approx 0.9621$ and $\max_{0 \le \sigma \le 1} \rho(A+\sigma(B-A)) \approx 0.999617$, where the maximum is achieved for $\sigma \approx 0.1266$. On the hand

$$A \cdot B = \begin{pmatrix} 0.3388 & 0.3675 \\ 0.6264 & 0.677 \end{pmatrix} \quad \text{and } \rho(A\cdot B) \approx 1.0166.$$

Note, that every $A + \sigma(B-A)$, $0 \le \sigma \le 1$ is positive. The condition $\rho(C_1 \cdot C_2) < 1$ for $C_1, C_2 \in \mathfrak{C}$ is necessary because

$$Z + \mathfrak{C}\cdot X \overset{\circ}{\subseteq} X \quad \text{implies} \quad Z + \mathfrak{C}\cdot Z + \mathfrak{C}\cdot\mathfrak{C}\cdot X \subseteq Z + \mathfrak{C}\cdot X \overset{\circ}{\subseteq} X$$

for $Z \in \mathbb{P}V\mathbb{R}$, $X \in \mathbb{P}V\mathbb{R}$, X compact and $\mathfrak{C} \in \mathbb{P}M\mathbb{R}$. Then

$$(Z + \mathfrak{C}\cdot Z) + (C_1 \cdot C_2)\cdot X \subseteq Z + \mathfrak{C}\cdot Z + \mathfrak{C}\cdot\mathfrak{C}\cdot X \overset{\circ}{\subseteq} X \quad \text{for every} \quad C_1, C_2 \in \mathfrak{C}$$

implying $\rho(C_1 \cdot C_2) < 1$ by theorem 5.

Finally we note the generalization of theorem 14 using ε-inflation. This is not clear, because there the inclusion in the convex union of all previous iterates is assumed.

Theorem 23. Let $Z \in \mathbb{P}V\mathbb{R}$, $\mathfrak{C} \in \mathbb{P}M\mathbb{R}$ and $\emptyset \neq X^0 \in \mathbb{P}V\mathbb{R}$, all $Z, \mathfrak{C}$ and X^0 being compact. Define $f : \mathbb{P}V\mathbb{R} \to \mathbb{P}V\mathbb{R}$ by $f(V) := Z + \mathfrak{C}\cdot V$ and let $g_i : \mathbb{P}V\mathbb{R} \to \mathbb{P}V\mathbb{R}$ with $V \in \mathbb{P}V\mathbb{R} \Rightarrow g_i(V) \supseteq V$ for $i \geq 0$. Define

$$X^{k+1} := g_{k+1}(f(X^k)) \quad \text{for} \quad 0 \leq k \in \mathbb{N}.$$

If then

$$f(X^m) \overset{\circ}{\subseteq} \bigcup_{i=0}^{m} X^i \quad \text{for some} \quad 0 \leq m \in \mathbb{N}, \tag{49}$$

then for every $z \in Z$ and for every $C \in \mathfrak{C}$ holds $\rho(C) < 1$ and there is one and only one $\hat{x} \in V\mathbb{R}$ with $z + C\cdot\hat{x} \in \hat{x}$. It is $\hat{x} = (I-C)^{-1} \in \bigcup_{i=0}^{m} X^i$.

Proof. Let $Y := \bigcup_{i=0}^{m} X^i$. We first proof by induction

$$f^k(X^{m-k+1}) \overset{\circ}{\subseteq} Y \quad \text{for} \quad 1 \leq k \leq m+1.$$

This is true for $k=1$ by assumption (49). Then

$$f^{k+1}(X^{m-k}) = f^k(f(X^{m-k})) \subseteq f^k(g_{m-k+1}(f(X^{m-k}))) = f^k(X^{m-k+1}) \overset{\circ}{\subseteq} Y. \tag{50}$$

Furthermore

$$f(Y) = \bigcup_{i=0}^{m} f(X^i) \subseteq f(X^m) \cup \bigcup_{i=0}^{m-1} g_{i+1}(f(X^i)) = f(X^m) \cup \bigcup_{i=1}^{m} X^i \subseteq Y, \tag{51}$$

implying $f^i(Y) \subseteq Y$ for $0 \leq i \in \mathbb{N}$ (but not yet $f^k(Y) \overset{\circ}{\subseteq} Y$ for some $k \in \mathbb{N}$). But

$$f^{m+1}(Y) = \bigcup_{i=0}^{m} f^{m+1}(X^i) = \bigcup_{k=1}^{m+1} f^{m+1-k}(f^k(X^{m-k+1})) \overset{\circ}{\subseteq} \bigcup_{k=1}^{m+1} f^{m+1-k}(Y) \subseteq Y$$

by (50) and (51). Y is compact because $Z, \mathfrak{C}$ and X^0 are. Therefore application of theorem 13 completes the proof. □

Especially interesting for practical applications is the following corollary.

Corollary 24. Let $Z \in \mathbb{P}V\mathbb{R}$, $\mathfrak{C} \in \mathbb{P}M\mathbb{R}$ and $\emptyset \neq X^0 \in \mathbb{P}V\mathbb{R}$, all $Z, \mathfrak{C}$ and X^0 being compact. Define $f : \mathbb{P}V\mathbb{R} \to \mathbb{P}V\mathbb{R}$ by $f(Y) := Z + \mathfrak{C}\cdot Y$ and let $g_i : \mathbb{P}V\mathbb{R} \to \mathbb{P}V\mathbb{R}$ with $V \in \mathbb{P}V\mathbb{R} \Rightarrow g_i(V) \supseteq V$ for $i \geq 0$. Define

$$X^{k+1} := g_{k+1}(f(X^k)) \quad \text{for} \quad 0 \leq k \in \mathbb{N}.$$

If then

$$f(X^m) \overset{\circ}{\subseteq} \bigcup_{i=0}^{m} X^i \quad \text{for some} \quad 0 \leq m \in \mathbb{N}, \tag{52}$$

then for every $z \in Z$ and for every $C \in \mathbb{C}$ holds $\rho(C) < 1$ and there is one and only one $\hat{x} \in V\mathbb{R}$ with $z + C\cdot\hat{x} = \hat{x}$. It is $\hat{x} = (I-C)^{-1} \in \bigcup_{i=0}^{m} X^i$.

Proof. Follows by replacing $\cup$ by $\underline{\cup}$ in the proof of theorem 23 because f is affine (see the remarks after corollary 15). □

The application of theorem 23 and corollary 24 to intervals of vectors are like shown before. Corollary 24 allows inclusions for iteration matrices C with $\rho(C) < 1$ but $\rho(|C|) \geq 1$. An example is

$$C := \begin{pmatrix} -10 & -9.4 \\ 10.1 & 9.5 \end{pmatrix} \quad \text{with} \quad \rho(C) = 0.6 \quad \text{and} \quad \rho(|C|) \approx 19.5.$$

For $f(X) := C\cdot X$ and $X^o := (1,-1)^T$ is $f(X^o) = (-0.6, 0.6)^T$. With an ε-inflation let $X^1 := ([-0.61,-0.59],[0.59,0.61])^T$. Then

$$f(X^1) = \begin{pmatrix} 0.166 \\ -0.164 \end{pmatrix} \underline{\cup} \begin{pmatrix} 0.354 \\ -0.354 \end{pmatrix} \underline{\cup} \begin{pmatrix} 0.366 \\ -0.366 \end{pmatrix} \underline{\cup} \begin{pmatrix} 0.554 \\ -0.556 \end{pmatrix} \subseteq X^o \,\underline{\cup}\, X^1 \tag{53}$$

proving $\rho(C) < 1$. As follows by theorem 6, $f(X^1) \overset{\circ}{\subseteq} X^o \,\underline{\cup}\, X^1$ cannot be satisfied. In practice, especially for larger n, condition (52) is difficult to verify. To compute the convex hull of the $X^o,\ldots,X^m$ is rather time consuming. On the other hand, $\Diamond(\bigcup_{i=0}^{m} X^i)$ resp. $O(\underline{\bigcup}_{i=0}^{m} X^i)$ can be computed very fast. Therefore one might wish to generalize corollary 24 in the way, that (52) could be replaced by

$$f(X^m) \overset{\circ}{\subseteq} \Diamond(\bigcup_{i=0}^{m} X^i) \quad \text{for some} \quad m \in \mathbb{N}. \tag{54}$$

Unfortunately this is not true. Consider the following example:

$$C := \begin{pmatrix} 0.26 & 0.76 \\ 0.76 & 0.26 \end{pmatrix} \quad \text{with} \quad \rho(C) = 1.02.$$

For $X^{k+1} := C\cdot X^k$ and $X^o := (1,0)^T$ is $X^1 = (0.26, 0.76)^T$ and $X^2 = (0.6452, 0.3952)^T$ with

$$X^2 \overset{\circ}{\subseteq} (X^o \,\underline{\cup}\, X^1) = \begin{pmatrix} [0.26,1] \\ [0,0.76] \end{pmatrix}.$$

The example shows, that (54) does not imply $\rho(C) < 1$. However, the contraction is very probable if (54) is satisfied as turned out in many practical examples. Therefore, in an algorithm, condition (54) could be tested and, for a final verification,

$$f(Y) \overset{\circ}{\subseteq} Y \quad \text{with } Y := \Diamond\left(\bigcup_{i=0}^{m} X^i\right)$$

has to be checked. This method is very effective and inexpensive.

4. Nonlinear systems of equations. In this chapter theorems are developed for the inclusion of the solution of general systems of nonlinear equations, the assumptions of which, again, are verifyable on computers. For this purpose we first linearize a given function $f : V\mathbb{R} \to V\mathbb{R}$, $f \in C^1$ locally. For $f' : V\mathbb{R} \to M\mathbb{R}$ being the Jacobian of f holds

$$f_i(x) = f_i(\tilde{x}) + f_i'(\tilde{x} + \theta_i(x - \tilde{x}))\cdot(x - \tilde{x}) \quad \text{for} \quad 1 \leq i \leq n, \quad 0 < \theta_i < 1, \tag{55}$$

where $f = (f_1,\ldots,f_n)^T$ and f_i' is the i-th row of the Jacobian f'. Defining $\mathbf{f}' : \mathbb{P}V\mathbb{R} \to \mathbb{P}V\mathbb{R}$ by

$$\mathbf{f}_{ij}(X) := \{f_{ij}'(x) \mid x \in X\} \quad \text{for} \quad X \in \mathbb{P}V\mathbb{R} \tag{56}$$

yields

$$f(x) \in f(\tilde{x}) + \mathbf{f}'(\tilde{x} \,\underline{\cup}\, x)\cdot(x - \tilde{x}) \quad \text{for} \quad \tilde{x}, x \in V\mathbb{R}. \tag{57}$$

Applying (57) to theorem 11 yields the following.

Theorem 25. Let $f : V\mathbb{R} \to V\mathbb{R}$ with $f \in C^1$, $\tilde{x} \in V\mathbb{R}$, $R \in M\mathbb{R}$ and $\emptyset \neq X \in \mathbb{P}V\mathbb{R}$ compact be given. If with $f' : \mathbb{P}V\mathbb{R} \to \mathbb{P}M\mathbb{R}$ defined in (56)

$$\tilde{x} - R\cdot f(\tilde{x}) + \{I - R\cdot\mathbf{f}'(\tilde{x} \,\underline{\cup}\, X)\}\cdot(X - \tilde{x}) \overset{\circ}{\subseteq} X, \tag{58}$$

Then R and every $M \in M\mathbb{R}$ with $M \in \mathbf{f}'(\tilde{x} \,\underline{\cup}\, X)$ is not singular and there is one and only one $\hat{x} \in X$ with $f(\hat{x}) = 0$. It is $\hat{x} \in \overset{\circ}{X}$.

Proof. Condition (58) implies

$$x - R\cdot\{f(\tilde{x}) + \mathbf{f}'(x \,\underline{\cup}\, X)\cdot(x - \tilde{x})\} \overset{\circ}{\subseteq} X \quad \text{for every } x \in X.$$

By (57) and the definition of $\mathbf{f}'$ this yields

$$\{x - R\cdot f(x) \mid x \in X\} \overset{\circ}{\subseteq} X. \tag{59}$$

By Brouwer's Fixed Point Theorem this implies the existence of an $\hat{x} \in X$ with $R \cdot f(\hat{x}) = 0$, where (59) shows $\hat{x} \in \overset{0}{X}$. By theorem 11 the matrix R and every matrix $M \in \mathbf{f}'(\tilde{x} \underline{\cup} X)$ is not singular implying $f(\hat{x}) = 0$. Suppose $f(\hat{y}) = 0$ for $\hat{y} \in X$. Then by the definition of $\mathbf{f}'$ and by (58) there is a $M \in \mathbf{f}'(\tilde{x} \underline{\cup} X)$ with

$$f(\hat{x}) = f(\hat{y}) + M(\hat{x} - \hat{y})$$

implying $M \cdot (\hat{x} - \hat{y}) = 0$ and therefore $\hat{x} = \hat{y}$. □

It has been noted in the literature (cf. [5], [14]) that the set of Jacobians defined in (56) can be weakened. Consider the following lemma.

Lemma 26. Let $f : V\mathbb{R} \to \mathbb{R}$ with $f \in C^1$ and $\tilde{x} \in V\mathbb{R}$. Then for every $x \in V\mathbb{R}$ there are $\theta_j \in \mathbb{R}$, $1 \le j \le n$ with $0 < \theta_j < 1$ and

$$f(x) = f(\tilde{x}) + \sum_{j=1}^{n} \{\frac{\partial f}{\partial x_j}(\tilde{x}_1,\ldots,\tilde{x}_{j-1},\tilde{x}_j + \theta_j(x_j-\tilde{x}_j),x_{j+1},\ldots,x_n)\cdot(x_j-\tilde{x}_j)\}. \quad (60)$$

Proof. Follows by straightforward calculation. □

Let $f : V\mathbb{R} \to V\mathbb{R}$ with $f \in C^1$ and $f = (f_1,\ldots,f_n)^T$. Then for $\tilde{x} \in V\mathbb{R}$ and the definition of $\mathbf{f}' : \mathbb{P}V\mathbb{R} \to \mathbb{P}V\mathbb{R}$ for $X \in \mathbb{I}V\mathbb{R}$ by

$$\mathbf{f}'_{ij}(X) := \{\frac{\partial f_i}{\partial x_j}(\tilde{x}_1,\ldots,\tilde{x}_{j-1},\tilde{x}_j \underline{\cup}\, x_j,x_{j+1},\ldots,x_n) \mid x_k \in X_k \text{ for } j \le k \le n\} \quad (61)$$

follows by lemma 26:

$$f(x) \in f(\tilde{x}) + \mathbf{f}'(X)\cdot(x - \tilde{x}). \quad (61)$$

Note, that definition (61) depends on $\tilde{x}$. Formula (62) implies that f' defined by (56) can be replaced by definition (61) in theorem 25. However, the uniqueness of the zero $\hat{x}$ of f in X cannot be proved like in theorem 25. There, for some $\hat{x},\hat{y} \in X$ with $f(\hat{x}) = f(\hat{y}) = 0$ the existence of a matrix $M \in \mathbf{f}'(X)$ was assumed with $f(\hat{x}) = f(\hat{y}) + M\cdot(\hat{x}-\hat{y})$. This need not to be true when using (61) instead of (56).

For our purposes, especially when applying Einzelschrittverfahren, another formulation of lemms 26 gives better results. Consider the following lemma.

Lemma 27. Let $f : V\mathbb{R} \to V\mathbb{R}$ with $f \in C^1$ and $\tilde{x} \in V\mathbb{R}$. Then for every $x \in V\mathbb{R}$ there are $\theta_{ij} \in \mathbb{R}$, $1 \le i,j \le n$ with $0 < \theta_{ij} < 1$ and

$$\mathbf{f}(x) = f(\tilde{x}) + M\cdot(x - \tilde{x}) \quad \text{with } M \in M\mathbb{R} \quad \text{and}$$
$$M_{ij} := \frac{\partial f_i}{\partial x_j}(x_1, x_2, \ldots, x_{j-1}, \tilde{x}_j + \theta_{ij}(x_j - \tilde{x}_j), \tilde{x}_{j+1}, \ldots, \tilde{x}_n). \tag{63}$$

Proof. Let i with $1 \leq i \leq n$ fixed but arbitrary. Then by the -dimensional Mean-Value Theorem

$$f_i(x_1,x_2,\ldots,x_n) = f_i(x_1,\ldots,x_{n-1},\tilde{x}_n) + \frac{\partial f_i}{\partial x_n}(x_1,\ldots,x_{n-1},\tilde{x}_n + \theta_{in}(x_n - \tilde{x}_n))\cdot$$
$$\cdot(x_n-\tilde{x}_n),$$
$$f_i(x_1,\ldots,x_{n-1},\tilde{x}_n) = f_i(x_1,\ldots,\tilde{x}_{n-1},\tilde{x}_n) + \frac{\partial f_i}{\partial x_{n-1}}(x_1,\ldots,x_{n-2},\tilde{x}_{n-1} + \theta_{i(n-1)}\cdot$$
$$\cdot(x_{n-1}-\tilde{x}_{n-1}),\tilde{x}_n)\cdot(x_{n-1}-\tilde{x}_{n-1}),\ldots,$$

$$f_i(x_1,\tilde{x}_2,\ldots,\tilde{x}_n) = f_i(\tilde{x}_1,\ldots,\tilde{x}_n) + \frac{\partial f_i}{\partial x_1}(\tilde{x}_1 + \theta_{i1}(x_1-\tilde{x}_1),\tilde{x}_2,\ldots,\tilde{x}_n))\cdot(x_1-\tilde{x}_1)$$

and therefore

$$f_i(x) = f_i(\tilde{x}) + \sum_{j=1}^{n} \frac{\partial f_i}{\partial x_j}(x_1,\ldots,x_{j-1},\tilde{x}_j + \theta_{ij}(x_j-\tilde{x}_j),\tilde{x}_{j+1},\ldots,\tilde{x}_n)\cdot(x_j-\tilde{x}_j)$$

demonstrating the lemma. □

Next the usage of the Einzelschrittverfahren for systems of nonlinear equations will be introduced. Before doing this the technique of computing an inclusion of the difference between an approximate solution and the true solution (like for linear systems theorem 18) will be introduced using definition (63) for **f'**.

Theorem 28. Let $f : V\mathbb{R} \to V\mathbb{R}$ with $f \in C^1$, $\tilde{x} \in V\mathbb{R}$, $R \in M\mathbb{R}$ and $X \in \mathbb{I}V\mathbb{R}$ be given. Let $\mathbf{f}' : \mathbb{I}V\mathbb{R} \to \mathbb{P}M\mathbb{R}$ be defined by

$$\mathbf{f}'_{ij}(Y) := \{\frac{\partial f_i}{\partial x_j}(x_1,\ldots,x_{j-1},\bar{x}_j,\tilde{x}_{j+1},\ldots,\tilde{x}_n) \mid x_i \in Y_i \text{ for } 1 \leq i < j,\ \bar{x}_j \in \tilde{x}_j \underline{\cup} Y_j\} \tag{64}$$

for $Y \in \mathbb{I}V\mathbb{R}$ and $1 \leq i,j \leq n$. If then

$$-R\cdot f(\tilde{x}) + \{I - R\cdot\mathbf{f}'(\tilde{x} + X)\}\cdot X \subsetneqq X, \tag{65}$$

then R and every matrix $M \in M\mathbb{R}$ with $M \in \mathbf{f}'(\tilde{x} + X)$ is not singular and there is an $\hat{x} \in \tilde{x} + X$ with $f(\hat{x}) = 0$.

Proof. By the definition (64) of **f'**, lemma 27 and (65) follows

$$\{x - R\cdot(f(\tilde{x}) + \mathbf{f}'(x + X)\cdot x) \mid x \in X\} \subsetneqq X$$

and therefore

$$\{x - R\cdot f(\tilde{x} + x) \mid x \in X\} \subsetneq X.$$

Therefore regarding theorem 6 the proof of theorem 25 can be applied.

□

Like in the case of systems of linear equations the inclusion of the difference of an approximate solution to the correct solution yields much sharper results than an inclusion of the solution itself. Next the application of the Einzelschrittverfahren is presented combined with the technique of including the residue with respect to an approximate solution.

Theorem 29. Let $f : V\mathbb{R} \to V\mathbb{R}$ with $f \in C^1$, $\tilde{x} \in V\mathbb{R}$, $R \in M\mathbb{R}$ and $x \in \mathbb{I}V\mathbb{R}$ be given. Let $\mathbf{f}' : \mathbb{I}V\mathbb{R} \to \mathbb{P}M\mathbb{R}$ be defined by

$$\mathbf{f}'_{ij}(V) := \{\frac{\partial f_i}{\partial x_j}(x_1,\ldots,x_{j-1},\bar{x}_j,\tilde{x}_{j+1},\ldots,\tilde{x}_n) \mid x_i \in V_i \text{ for } 1 \le i < j, \bar{x}_j \in \tilde{x}_j \underline{\cup} V_j\} \quad (66)$$

for $V \in \mathbb{I}V\mathbb{R}$ and $1 \le i,j \le n$. Define recursively $Y \in \mathbb{I}V\mathbb{R}$ by

$$Y_i := \lozenge(-R\cdot f(\tilde{x}) + \{I - R\cdot\mathbf{f}'(\tilde{x} + Z)\}\cdot Z)_i \quad \text{for} \quad 1 \le i \le n \qquad (67)$$

where $Z_i := (Y_1,\ldots,Y_{i-1},X_i,\ldots,X_n)^T \in \mathbb{I}V\mathbb{R}$. If then

$$Y_i \subsetneq X_i \quad \text{for} \quad 1 \le i \le n, \qquad (68)$$

then the matrix R and every matrix $M \in M\mathbb{R}$ with $M \in \mathbf{f}'(\tilde{x} + Y)$ is not singular and there is an $\hat{x} \in \tilde{x} + Y$ with $f(\hat{x}) = 0$. Moreover with

$$W^o := Y; \quad W^{k+1} := \lozenge(-R\cdot f(\tilde{x}) + \{I - R\cdot\mathbf{f}'(\tilde{x} + W^k)\}\cdot W^k)$$

holds

$$\hat{x} \in \bigcap_{k\ge 0} W^k .$$

Proof. By (67), the definition of the Z_i and (68) holds

$$\lozenge(-R\cdot f(\tilde{x}) + \{I - R\cdot\mathbf{f}'(\tilde{x} + Y)\}\cdot X) \subsetneq X \quad \text{using Einzelschrittverfahren} \qquad (69)$$

as described in theorem 18 with the rounding $\lozenge$ defined in chapter 1. Therefore by theorem 12 the matrix R and every matrix $M \in M\mathbb{R}$ with $M \in \mathbf{f}'(\tilde{x} + Y)$ is not singular. Furthermore by (67),(69) and theorem 12

$$\{x - R\cdot(f(\tilde{x}) + f'(\tilde{x} + Y)\cdot x) \mid x \in Y\} \subseteq Y$$

and therefore by definition (66) and lemma 27

$$\{x - R\cdot f(\tilde{x} + x) \mid x \in Y\} \subseteq Y. \tag{70}$$

By Brouwer's Fixed Theorem there is an $\hat{y} \in Y$ with $\hat{y} - R\cdot f(\tilde{x} + \hat{y}) = \hat{y}$ implying $f(\tilde{x} + \hat{y}) = 0$ by the non-singularity of R. With $\hat{x} := \tilde{x} + \hat{y}$ the proof is complete observing (69) and (70). □

5. Implementation and examples. Following some implementation hints and numerical examples will be given. Implementation details will be given for systems of linear equations; they apply for systems of nonlinear equations as well.

To calculate an inclusion of a linear system $Ax = b$ first an approximate inverse R is required according to theorem 18. Note, that R can be replaced by LU from an LU-decomposition. The success of the algorithm, i.e. whether an inclusion will be computed or not, depends highly on R. In fact, as theorem 22 shows, an inclusion will be found if and only if $I - RA$ is convergent, However, it is important to note, that the user does not have to know a priori, whether R or A is not singular, he does not have to know in advance, whether the spectral radius of $I - RA$ is less than one. This will be demonstrated by the algorithm automatically a posteriori.

In theorem 18, also an approximate solution $\tilde{x}$ is required. The number of iterations (3. in (41)) necessary depends, of course, on the quality of $\tilde{x}$. However, again no additional information on $\tilde{x}$ such as $\|\hat{x} - \tilde{x}\|$ is required.

Suppose, R and $\tilde{x}$ are given (e.g. computed using some traditional method, cf.[11],[42],[43]), then an algorithm based on theorem 17, on the third iteration in (41) and the ε-inflation would be the following:

$$X := \lozenge(b - A\tilde{x});\; Z := R \odot X;\; C := \lozenge(I - R\cdot A);\; X := Z;\; k = 0;$$

$$\underline{\text{repeat}}\;\; k := k+1;\; Y := X \circ \varepsilon;\; X := Z \oplus C \odot Y \;\; \underline{\text{until}} \;\; X \overset{\circ}{\subseteq} Y \;\; \underline{\text{or}} \;\; k > 15;$$

Algorithm 1. Traditional way.

$\lozenge : \mathbb{P}V\mathbb{R} \to \mathbb{I}VG$ resp. $\lozenge : \mathbb{P}M\mathbb{R} \to \mathbb{I}MG$ is any isotone rounding, i.e. $A \subseteq \lozenge A$. As has been pointed out before, it is of utmost importance to compute the residuals $b - A\tilde{x}$ and $I - RA$

with one rounding. Preferably, a precise scalar product with maximum accuracy introducing only one rounding error is used (cf.[7],[8]). In this case the rounding O is best possible and therefore equal to ◇.
X,Y and Z are interval vectors and C is an interval matrix; therefore the additional storage needed is $6n + 2n^2 + O(1)$. Next, the Einzelschrittverfahren can be used according to theorem 18 yielding the following algorithm.

$X := O(b - A\tilde{x})$; $Z := R \odot X$; $C := O(I - R \cdot A)$; $X := Z$; $k := 0$;
<u>repeat</u> $k := k + 1$; $X := X \circ \varepsilon$; incl := true;
  <u>for</u> $i := 1$ <u>to</u> n <u>do</u> {$Q := Z_i \oplus C_i \odot X$; <u>if</u> $Q \not\subseteq \overset{\circ}{X}_i$ <u>then</u> incl:= false;
    $X_i := A$}
<u>until</u> incl <u>or</u> $k > 15$;

<u>Algorithm 2</u>. Einzelschrittverfahren

Here C_i denotes the i-th row of C, X_i the i-th component of X. Q is an interval and the additional storage needed is $4n + 2n^2 + O(1)$.

In order to avoid the additional $O(n^2)$ storage consider the following. When using interval vectors as subsets of $V\mathbb{R}$ it has been shown in theorem 12 that

$$Z + \mathfrak{C} \cdot X \subsetneq X \quad \text{for} \quad X, Z \in \mathbb{I}V\mathbb{R} \; ; \; \mathfrak{C} \in \mathbb{I}M\mathbb{R} \tag{71}$$

already implies $\rho(|C|) < 1$ for every matrix $C \in \mathfrak{C}$. Therefore, instead of (71),

$$|Z| + |\mathfrak{C}| \cdot |X| < |X| \tag{72}$$

can be checked. Necessary and sufficient conditions for an iteration using (72) to stop have been given in [39].

A significant amount of storage can be saved because only the absolute value of X,Z and C are needed. Therefore the additional storage needed essentially halves to $3n + n^2 + O(1)$. Moreover, now the matrix C, which is a point matrix, can be computed in the storage of R. This is possible because $C := I - RA$ can be computed rowwise, storing the intermediate (row) result somewhere. After the first row of C has been computed, the first row of R is no longer needed. This leads to the following algorithm.

$[x,y] := \Diamond(b - A\tilde{x})$; $z := \Delta(|R\cdot[x,y]|)$;

<u>for</u> $i := 1$ <u>to</u> n <u>do</u> $\{x := R_i;\ R_i := \Delta(|I_i - x\cdot A|)\}$; $x := z$; $k := 0$;

<u>repeat</u> $k := k+1$; $x := x \triangle\!\!\!\!\!\circ\ \varepsilon$; incl := true;

<u>for</u> $i:=1$ <u>to</u> n <u>do</u> $\{q := z_i \triangle\!\!\!\!+ R_i \triangle x$; <u>if</u> $q \geq x_i$ <u>then</u> incl := false;
$x_i := q\}$

<u>until</u> incl <u>or</u> $k > 15$;

<u>Algorithm 3</u>. Checking (72) with Einzelschrittverfahren

Algorithms 1 and 2 imply $\hat{x} \in \tilde{x} \oplus X$ ($\hat{x} := A^{-1}\cdot b$) whereas algorithm 3 implies $\hat{x} \in \tilde{x} \oplus [-x,x]$. The validity of algorithm 3 follows by

$$|Z| + |\mathfrak{C}|\cdot|X| < |X| \Rightarrow -|X| < -|Z| - |\mathfrak{C}|\cdot|X| \leq |Z| + |\mathfrak{C}|\cdot|X| < |X| \Rightarrow$$
$$\Rightarrow [-|Z|,|Z|] + [-|\mathfrak{C}|,|\mathfrak{C}|]\cdot[-|X|,|X|] \overset{\circ}{\subseteq} [-|X|,|X|] \Rightarrow$$
$$\Rightarrow Z + \mathfrak{C}\cdot[-|X|,|X|] \overset{\circ}{\subseteq} [-|X|,|X|].$$

All three algorithms verify the non-singularity of R and A and therefore the unique solvability of the linear system $Ax = b$. The additional storage required for algorithms 1 to 3 is:

algorithm	1	2	3
storage	$2n^2 + 6n + O(1)$	$2n^2 + 4n + O(1)$	$3n + O(1)$

For algorithm 3 the computing reduces as well.

The non-singularity of R and A could also be demonstrated by showing $\| I-RA \| < 1$ for some norm $\| \cdot \| : M\mathbb{R} \to \mathbb{R}$. On the computer, $\| \Delta(|I-RA|) \| < 1$ would be verified. In the following some numerical examples are shown where $\| I-RA \| \geq 1$ for a number of norms but nevertheless an algorithm based on the inclusion theory does verify the contraction and delivers sharp bounds for the solution of the linear system.

The following examples are calculated on an IBM S/370 in single precision (~ 7.5 decimal digits) and double precision (~16.5 decimal digits). The approximate inverse R is computed using Gauß-Jordan algorithm. As examples consider (n is the number of rows)

Hilbert* - matrices $H^*_{ij} := \dfrac{\mathrm{lcm}(1,2,\ldots,2n-1)}{i+j-1}$

Pascal - matrices $P_{ij} := \binom{i+j}{j}$

Pascal* - matrices $\quad P^*_{ij} := \binom{i+j-1}{j}$

Zielke - matrices $\quad Z_{ij} := \dfrac{\binom{n+i-1}{i-1} \cdot n \cdot \binom{n-1}{n-j}}{i+j-1}$ (cf.[50])

$S_{ij}(q) := 1 - q \cdot r_{ij}$ where $r_{ij} \in [0,1]$ randomly.

All matrices except the last have integer entries. Where the precision in use does not suffice to store an entry precisely, the entry rounded to nearest is used.

In the following tables the matrix type and the dimension n is listed. In the column $\| I-RA \|$ the minimum value of sum-, maximum- and Frobenius-norm is listed. k is the number of interval iterations and "verified digits" is the minimum number of decimal digits coinciding of the bounds for the components of the solution using algorithm 3 listed above. In all cases a linear system with the depicted matrix and right hand side $(1,\dots,1)^T$ is solved. First the single precision results are displayed.

matrix	n	$\| I-RA \|$	k	verified digits
H	7	1.7	2	7.5
P	8	1.2	1	7.5
	9	38	2	7.3
P*	9	3.5	2	7.4
Z	7	1.6	3	7.3
$S(10^{-5})$	25	0.74	2	7.4

Table 1. Single precision results (~ 7.5 decimal digits)

As can be seen the number of verified decimal digits a widely independent on the condition of the matrix. This is due to the technique not to compute an inclusion of the solution itself but of the difference of the true solution $\hat{x}$ to an approximate solution $\tilde{x}$. Furthermore, the convergence of I - RA can be shown even if norm estimates fail to show $\rho(I-RA) < 1$. In the next table there are more extreme examples in this respect.

matrix	n	$\|I-RA\|$	k	verified digits
P	20	11	1	16.5
	22	210	1	16.5
	24	670	1	16.5
	26	280000	1	16.5
$S(10^{-3})$	50	0.02	1	16.5
	100	0.03	1	16.3
	200	0.49	2	16.4

Table 2. Double precision results (~16.5 decimal digits)

The Pascal 26×26 matrix shows the extreme example where the spectral radius of I-RA is estimated by 280000 where in fact the new methods show $\rho(I\text{-}RA) < 1$. Also for larger dimensions the inclusions behave very stable.

6. Conclusion. In the preceding chapters the theoretical foundations of the inclusion theory were extended and new proofs were given without using Brouwer's Fixed Point Theorem. Some hints on the implementation were also provided. In [38] the theoretical background and corresponding algorithms are given for other numerical problems such as linear systems with band matrix, symmetric matrix or sparse matrix, for over - and underdetermined linear systems, evaluation and zeros of polynomials, algebraic eigenvalue problems, linear, quadratic and convex programming problems, evaluation of arithmetic expressions and others. Algorithms corresponding to a number of those problems are implemented in the IBM Program Product ACRITH, which is available since March 1984 and with a second Release since early 1985 (cf.[51]).

The key property of the new algorithms is that the verification of the validity of the result is performed automatically by the computer without any effort on the part of the user. The verification includes the existence and uniqueness of a solution within the computed bounds. The input data may be real point or interval data as well as complex point or interval data. Especially if the data is afflicted with tolerances the verification process is of great help. In this case it is verified that any problem within the tolerances is solvable and the solution of any of the (infinitely many) problems within the tolerances is enclosed within the calculated inclusion interval.

The computing time is of the order of a comparable floating-point algorithm (e.g. Gaussian elimination in case of general linear systems with full matrix), the latter, of course, without the verification of the result.

The computed bounds are of high accuracy, i.e. the difference of the left and right bound of the inclusion of every component is of the order of the relative rounding error unit. By our experience, very often the inclusions are of least significant bit accuracy, i.e. the left and right bound of the inclusion of every component are adjacent floating-point numbers.

7. References

[1] Abbott, J.P., Brent, R.P. (1975). Fast Local Convergence with Single and Multistep Methods for Nonlinear Equations, Austr. Math. Soc. 19 (series B), 173-199.

[2] Alefeld, G., Intervallrechnung über den komplexen Zahlen und einige Anwendungen. Dissertation, Universität Karlsruhe, 1968.

[3] Alefeld, G. and Herzberger, J., "Einführung in die Intervallrechnung". Reihe Informatik, 12. Wissenschaftsverlag des Bibliographischen Instituts Mannheim, 1974.

[4] Alefeld, G. and Herzberger, J., "Introduction to Interval Analysis", Academic Press, New York (1983).

[5] Alefeld, G. (1979). Intervallanalytische Methoden bei nichtlinearen Gleichungen. In "Jahrbuch Überblicke Mathematik 1979", B.I. Verlag, Zürich.

[6] Bauer, F.L. and Samelson, K. Optimale Rechengenauigkeit bei Rechenanlagen mit gleitendem Komma, Z. Angew. Math. Phys. 4, 312-316 (1953).

[7] Bohlender, G., Floating-point computation of functions with maximum accuracy. IEEE Trans. Compute. C-26, No. 7, 621-632 (1977).

[8] Bohlender, G., Genaue Summation von Gleitkommazahlen, Computing Suppl. 1, 21-32 (1977).

[9] Collatz, L., "Funktionalanalysis und Numerische Mathematik," Springer-Verlag, Berlin and New York, 1968.

[10] Coonan, J., et al. (1979). A proposed standard for floating-point arithmetic, SIGNUM Newsletter.

[11] Forsythe, G.E., and Moler, C.B., "Computer Solution of Linear Algebraic Systems", Prentice-Hall, Englewood Cliffs, New Jersey, 1967.

[12] Forsythe, G.E., Pitfalls in computation, or why a math book isn't enough, Tech.Rep. No. CS147, pp. 1-43. Computer Science Department, Stanford University, Stanford,California, 1970.

[13] Haas, H.Ch., Implementierung der komplexen Gleitkommaarithmetik mit maximaler Genauigkeit. Diplomarbeit, Institut für Angewandte Mathematik, Universität Karlsruhe, 1975.

[14] Hansen, E., Interval Arithmetic in Matrix Computations, Part 1. SIAM J. Numer. Anal. 2, 308-320 (1965), Part II. SIAM J. Numer. Anal. 4, 1-9 (1967).

[15] Heuser, H. (1967). Funktionalanalysis. Mathematische Leitfäden, G.B. Teubner, Stuttgart.

[16] Kahan, W. and Parlett, B.N., Können Sie sich auf Ihren Rechner verlassen?, "Jahrbuch Überblicke Mathematik 1978". Wissenschaftsverlag des Bibliographischen Instituts Mannheim, pp. 199-216, (1978).

[17] Kaucher, E., Rump, S.M. (1982). E-methods for Fixed Point Equations f(x) = x, Computing 28, p.31-42.

[18] Kaucher, E., Miranker, W.L., "Self-Validating Numerics of Function Space Problems", Academic Press, New York (1984).

[19] INTEL 12 1586-001. (1980). The 8086 Family User's Manual, Numeric Supplement.

[20] Köberl, D. (1980). The Solution of Non-linear Equations by the Computation of Fixed Points with a Modification of Sandwich Method, Computing, 25, 175-178.

[21] Krawczyk, R. (1969). Newton-Algorithmen zur Bestimmung von Nullstellen mit Fehlerschranken, Computing, 4, 187-220.

[22] Knuth, D., "The Art of Computer Programming", Vol. 2 Addison-Wesley, Reading, Massachusetts, 1962.

[23] Kulisch, U., An axiomatic approach to rounded computations, TS Report No. 1020, Mathematics Research Center, University of Wisconsin, Madison, Wisconsin, 1969, and Numer. Math, 19, 1-17 (1971).

[24] Kulisch, U., Grundzüge der Intervallrechnung, Überblicke Mathematik 2, Bibliographisches Institut Mannheim, 51-98 (1969).

[25] Kulisch, U., Formalization and implementation of floating-point arithmetic, Computing 14, 323-348 (1975).

[26] Kulisch, U., "Grundlagen des Numerischen Rechnens-Mathematische Begründung der Rechnerarithmetik", Reihe Informatik, 19. Wissenschaftsverlag des Bibliographischen Instituts Mannheim, 1976.

[27] Kulisch, U., Ein Konzept für eine allgemeine Theorie der Rechnerarithmetik, Computing Suppl. 1, 95-105 (1977).

[28] Kulisch, U., Miranker, W.L. (1981). Computer Arithmetic in Theory and Practise. Academic Press, New York.

[29] Kulisch, U.W., Miranker, W.L. (eds.): A New Approach to Scientific Computation, Academic Press, New York, 1983.

[30] Moore, R.E., "Interval Analysis". Prentice-Hall, Englewood Cliffs, New Jersey, 1966.

[31] Moore, R.E. (1977). A Test for Existence of Solutions for Non-Linear Systems, SIAM J. Numer. Anal., 4.

[32] Moré, J.J., Cosnard, M.Y. (1979). Numerical Solution of Non-Linear Equations. ACM Trans. on Math. Software, Vol. 5, No. 1, 64-85.

[33] Ortega, J.M., Reinboldt, W.C. (1970). Iterative Solution of Non-linear Equations in several Variables. Academic Press, New York-San Francisco-London.

[34] Perron, O., "Irrationalzahlen". de Gruyter, Berlin, 1960.

[35] Rall, L.B. (1981). Mean value and Taylor forms in interval analysis, SIAM J. Math. Anal. 14, No. 2 (1983).

[36] Reinsch, Ch., Die Behandlung von Rundungsfehlern in der Numerischen Analysis, "Jahrbuch Überblicke Mathematik 1979", Wissenschaftsverlag des Bibliographischen Instituts Mannheim, 43-62 (1979).

[37] Rump, S.M. (1980). Kleine Fehlerschranken bei Matrixproblemen, Dissertation, Universität Karlsruhe.

[38] Rump, S.M. (1983). Solving Algebraic Problems with High Accuracy, Habilitationsschrift, in Kulisch/Miranker: A New Approach to Scientific Computation, Academic Press, New York.

[39] Rump, S.M. (1982). Solving Non-linear Systems with Least Significant Bit Accuracy, Computing 29, 183-200.

[40] Rump, S.M. (1984). Solution of Linear and Nonlinear Algebraic Problems with Sharp, Guaranteed Bounds, Computing Suppl. 5, 147-168.

[41] Rump, S.M. and Kaucher, E., Small bounds for the solution of systems of linear equations, Computing Suppl. 2, 157-164 (1980).

[42] Stoer, J. (1972). Einführung in die Numerische Mathematik I. Heidelberger Taschenbücher, Band 105, Springer-Verlag, Berlin-Heidelberg-New York.

[43] Stoer, J., Bulirsch, R. (1973). Einführung in die Numerische Mathematik II. Heidelberger Taschenbücher, Band 114, Springer-Verlag, Berlin-Heidelberg-New York.

[44] Ullrich, Ch., Zur Konstruktion komplexer Kreisarithmetiken Computing Suppl. 1, 135-150 (1977).

[45] Varga, R.S. (1962). Matrix Iterative Analysis. Prentice-Hall, Englewood Cliffs, New Jersey.

[46] Walter, W. (1970). Differential and integral Inequalities. Berlin-Heidelberg-New York: Springer.

[47] Wilkinson, J.H., "Rounding Errors in Algebraic Processes". Prentice-Hall, Englewood Cliffs, New Jersey, 1963.

[48] Wongwises, P. Experimentelle Untersuchungen zur numerischen Auflösung von linearen Gleichungssystemen mit Fehlerfassung, Interner Bericht 75/1, Institut für Praktische Mathematik, Universität Karlsruhe.

[49] Yohe, J.M., Roundings in floating-point arithmetic, IEEE Trans. Comput. C.12 No. 6, 577-586 (1973).

[50] Zielke, R., Algol-Katalog Matrizenrechnung, Oldenburg Verlag, München, Wien (1972).

[51] ACRITH High-Accuracy Arithmetic Subroutine Library: General Information Manual, IBM Publications, GC33-6163, (1985).

Acknowledgement: The author wants to thank his students of the summer lecture 1985 for several helpful comments.

Address of the author:

Priv.-Doz. Dr. Siegfried M. Rump
IBM Development and Research
Schönaicher-Straße 220

D-7030 Böblingen
Federal Republic of Germany

ACCURATE ELLIPTIC DIFFERENTIAL EQUATION SOLVER

W. F. Ames and R. C. Nicklas
School of Mathematics
Georgia Institute of Technology
Atlanta, GA 30332

ABSTRACT

This report describes accurate numerical methods for general elliptic problems defined on rectangular domains with boundary conditions of the third kind. The continuous model is discretized using finite differences and the resulting system of linear algebraic equations is solved iteratively. Several iterative algorithms are implemented in both interval, using IBM's ACRITH, and point arithmetic. Problems are exhibited for which the point algorithms perform poorly but the corresponding interval procedures do not.

INTRODUCTION

Many problems in engineering and science are best described by partial differential equations (PDE). Due to the complexity of most natural phenomena, the resulting partial differential equation (or system of partial differential equations) is often analytically intractable. If predictions are to be made, the mathematical model must either be simplified or a numerical approach to the problem taken. In an effort to obtain a solution in closed form, a model may be over simplified to the point of uselessness. For example, many natural phenomena are inherently nonlinear and simply will not appear in linearized models. Even if a closed form expression for the solution of a PDE is obtained, the result may be so complicated that a study of its properties would be difficult.

The numerical approach to modeling involves the use of a digital computer to solve the relevant partial differential equations. To this end, the mathematical model must be transformed so that the solution process requires only the arithmetic operations of addition, subtraction, multiplication, and division. This transformation from a continuous model to one capable of solution on a digital computer is known as discretization. The two most widely employed methods of discretization are the finite difference method and the finite element method. The class of problems treated in this report will be transformed from continuous systems to discrete systems using finite difference methods.

In their work on the modeling of physical systems, J. von Neumann and H. H. Goldstine identified the following four major sources of error.

1. Modeling Errors
   If a mathematical model is derived from over simplified or erroneous assumptions, the modeling errors are introduced. These errors are the responsibility of the scientist or engineer, who typically possesses a certain amount of physical intuition for the problem at hand.
2. Measurement Errors
   Most descriptions of physical systems will require experimentally determined data as input. The error in a mathematical model due to uncertain input data is called measurement error.
3. Truncation Errors
   The mathematical description of a physical system almost always involves a continuum or some sort of infinite process. Approximation methods can deal with only a finite number of terms in an infinite series, a finite number of iterations in an iterative process, etc. Errors introduced in this context are called truncation errors.
4. Roundoff Errors
   When an approximation method is implemented on a digital computer, the problem of finite word length arises. The set of real numbers a given digital computer can manipulate is finite and in addition this set is not closed under any of the arithmetic operations. Errors introduced because of the resulting loss of digits are called roundoff errors.

The treatment of truncation and roundoff errors falls within the domain of numerical analysis. The rate at which the error in a finite difference approximation vanishes as the mesh is refined is proportional to the spacing of the mesh to some power. Thus, truncation errors may be reduced by either refining the mesh or using a higher order finite difference approximation. Roundoff errors have been traditionally treated by carrying out calculations in double or even quadruple precision. Even with extended word lengths, roundoff error can creep in to destroy the validity of a computation. Since computers capable of greater than 100 million floating point operations are now available, traditional computer arithmetic, along with its tedious error analyses, must be supplanted by more reliable methods.

Recent research activity in scientific computation has culminated in commercially available scientific software based on interval arithmetic. While the concept of interval numbers is not new (see Moore [10]), a complete formal mathematical description of the number spaces and their attendent arithmetic that arise when numerical calculations are carried out on digital computers has only recently been published

(Kulisch and Miranker [1]). The new software systems based on these mathematical researches provide for automatic roundoff error control. Guaranteed bounds on the true solution are supplied with full machine accuracy, and the computer simultaneously verifies the existence and uniqueness of the solution within the computed bounds. These new methods have been called E-methods, from the German Existenz (existence) Eindeutigkeit (uniqueness), and Einschliesung (inclusion). See the collection of papers contained in Kulisch and Miranker [1,2] for more information on the E-methods.

The numerical experiments described in this report were carried out using subroutines from the IBM ACRITH library of interval routines. The programs were executed on an IBM 4361 computer, a machine which implements in hardware the directed rounding operations that are essential to the new interval arithmetic. For a description of the ACRITH software, see IBM [3,4].

THE CONTINUOUS PROBLEM

Elliptic partial differential equations characterize distributed parameter systems whose fields are purely reservoirs of potential, or purely reservoirs of flux. In addition, fields containing dissipation as well as reservoir elements are described by elliptic equations once the steady state has been reached. Elliptic partial differential equations serve as mathematical models in the fields of electrostatics, irrotational incompressible fluid flow, steady state heat conduction, fission reactor theory, mechanics, diffraction theory, microwave wave guides and cavity resonators, and gravitation, among others (Samarski and Tychonov [5]). The elliptic partial differential equation which models a physical system usually holds on a simple bounded region and is accompanied by appropriate boundary conditions. An exception to this is the exterior boundary value problem, where the PDE holds on an unbounded region, the solution is to satisfy prescribed boundary conditions on a closed curve, and (for uniqueness) the solution vanishes at infinity (see Samarski and Tychonov [5], p. 265). The exterior BVP arises naturally in fluid flow problems.

Let Ω be a bounded plane region with boundary Γ. The general elliptic boundary value problem is to find a function $u(x,y)$ defined and continuous on $\Omega \cup \Gamma$ and twice continuously differentiable in Ω that satisfies the linear second order partial differential equation

$$Lu := A(x,y)u_{xx} + C(x,y)u_{yy} + D(x,y)u_x + E(x,y)u_y + F(x,y)u$$
$$= G(x,y)$$

in Ω. A, C, D, E, F, and G are taken to be analytic functions of x and y in Ω. Since L is an elliptic operator, the coefficient functions A and C must have the same sign. Without loss of generality, A and C may be taken to be positive and $F \leq 0$.

On the boundary, u is required to satisfy boundary conditions of the third kind:

$$\alpha(x,y)u(x,y) + \beta(x,y)\frac{\partial u}{\partial n}(x,y) = g(x,y).$$

Here g(x,y) is a given function that is continuous on Γ. At each point on the boundary, a linear combination of the field variable and its normal derivative is specified. Thus, Dirichlet and Neumann boundary conditions are special cases of boundary conditions of the third kind.

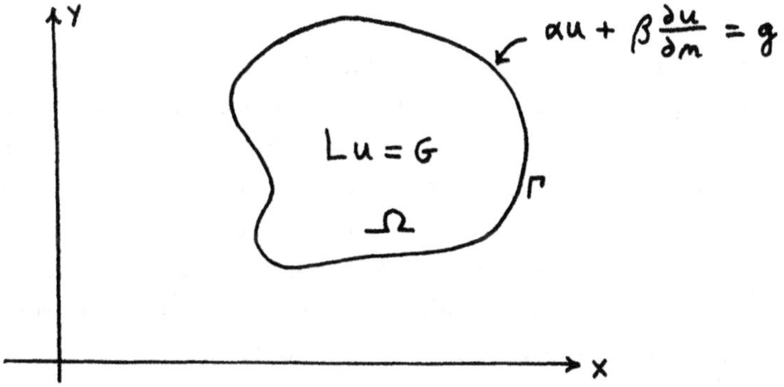

Figure 1. Elliptic Boundary Value Problem.

This report treats the following class of two dimensional elliptic problems:

$$Lu := A(x,y)u_{xx} + C(x,y)u_{yy} + D(x,y)u_x + E(x,y)u_y + F(x,y)u$$
$$= G(x,y) \quad \text{in } \Omega,$$

$$\Omega = \{(x,y): a < x < b \text{ and } c < y < d\},$$

subject to boundary condition of the third kind:

$$\text{left:} \quad B_1u := \alpha_0(y)u_x(a,y) + \beta_0(y)u(a,y) = \gamma_0(y),$$
$$\text{right:} \quad B_2u := \alpha_1(y)u_x(b,y) + \beta_1(y)u(b,y) = \gamma_1(y),$$
$$\text{bottom:} \quad B_3u := \delta_0(x)u_y(x,c) + \varepsilon_0(x)u(x,c) = \lambda_0(x),$$
$$\text{top:} \quad B_4u := \delta_1(x)u_y(x,d) + \varepsilon_0(x)u(x,d) = \lambda_1(x).$$

B_1, B_2, B_3, and B_4 are linear boundary operators.

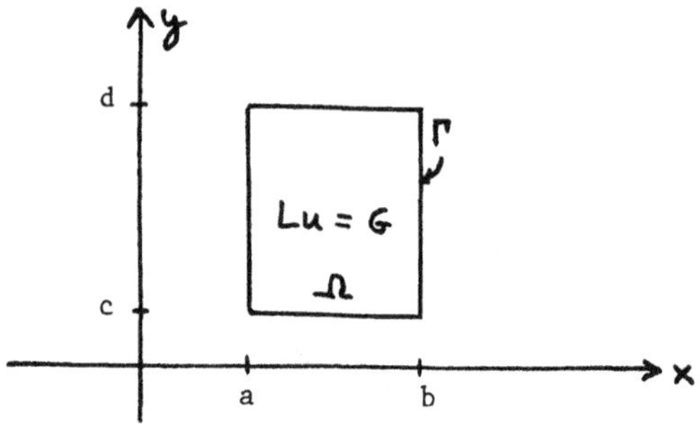

Figure 2. Elliptic BVP on a Rectangular Domain.

DISCRETIZATION

The Computational Mesh

The first step in discretizing an elliptic BV problem is to superimpose on the region Ω a finite set of discrete points, referred to as mesh or grid points. At each of the mesh points an algebraic equation approximating the partial differential equation or the boundary conditions is applied. The approximate solution to the boundary value problem is then given by the solution of the resulting system of algebraic equations. The unknown variables in this algebraic system are of course the approximate field values at the mesh points. Approximate values of the field at nonmesh points may be computed using a suitable interpolation process.

Since L is a linear operator, as are the boundary operators, the corresponding algebraic equations will be linear as well. Thus, the computation of an approximate solution to an elliptic boundary value problem reduces to solving a system of linear algebraic equations.

For the simple rectangular regions treated in this report, the obvious choice of mesh is a cartesian one. To ease the programming effort, the numerical experiments described in this report were carried out with a uniform mesh -- the mesh spacing in the vertical and horizontal directions is the same. The following figure illustrates the rectangular mesh superimposed on the region.

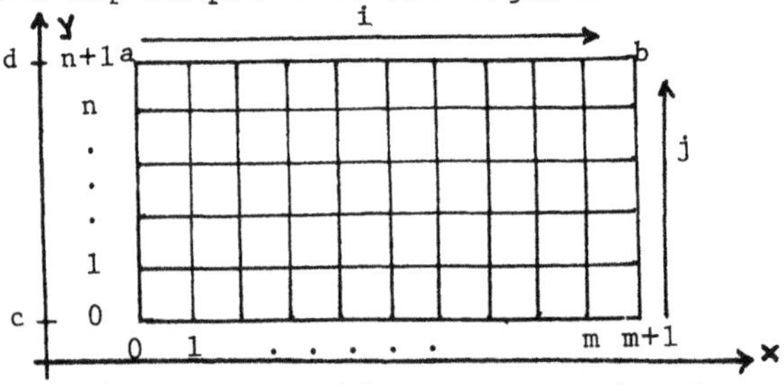

Figure 3. Uniform Computational Mesh.

The x-coordinates of the field points are labeled x_i and given by

$$x_i = a + ih,$$

where h is the uniform mesh spacing and the index i runs from 0 to m+1. The y-coordinates of the field points are denoted by y_j and given by

$$y_j = c + jh,$$

where the index j runs from 0 to n+1.

Note: Since the mesh spacing is uniform, m (the number of interior field points in the horizontal direction), n (the number of interior field points in the vertical direction), and h must satisfy

$$(b-a)/(m+1) = h = (d-c)/(n+1).$$

There are mn interior mesh points and 2m + 2n + 4 boundary mesh points. In order to handle the most general boundary conditions, those of the third kind, all mn + 2(m+n) + 4 = (m+2)(n+2) approximate field values (i.e., the approximate value of u at the grid points) will be treated as unknown variables. The next two sections describe how an equal number of linear equations in these unknown field variables are assembled.

Discretization of the Partial Differential Equation

Having established a computational mesh, the next step in the discretization process is to obtain finite difference approximations to the elliptic operator L at each of the interior mesh points as well as approximations to the boundary operators at the boundary grid points. The following notation is employed: $u_{ij} = u(x_i,y_j)$ is the exact value of the field at the mesh point (x_i,y_j) and U_{ij} is the discrete approximation to the field value at (x_i,y_j).

The following central differences are employed in the finite difference approximation to the operator L (Ames [6,7])

$$u_x(x_i,y_j) = (u_{i+1,j} - u_{i-1,j})/2h + O(h^2)$$
$$u_y(x_i,y_j) = (u_{i,j+1} - u_{i,j-1})/2h + O(h^2)$$
$$u_{xx}(x_i,y_j) = (u_{i+1,j} - 2u_{i,j} + u_{i-1,j})/h^2 + O(h^2)$$
$$u_{yy}(x_i,y_j) = (u_{i,j+1} - 2u_{i,j} + u_{i,j-1})/h^2 + O(h^2).$$

These second order accurate approximations, when inserted into the partial differential equation, yields

$$\beta_1 U_{i+1,j} + \beta_2 U_{i-1,j} + \beta_3 U_{i,j+1} + \beta_4 U_{i,j-1} - \beta_0 U_{i,j} = h^2 G_{i,j}$$

where

$$\beta_1 = A_{i,j} + 0.5hD_{i,j}$$

$$\beta_2 = A_{i,j} - 0.5hD_{i,j}$$
$$\beta_3 = C_{i,j} + 0.5hE_{i,j}$$
$$\beta_4 = C_{i,j} - 0.5hE_{i,j}$$
$$\beta_0 = 2(A_{i,j} + C_{ij} - 0.5h^2F_{ij})$$

and $i = 1,2,\ldots,m$, $j = 1,2,\ldots,n$. All the β_i will be positive if h is chosen so small that

$$0 < h < \min\left\{\frac{2A_{ij}}{|D_{ij}|}, \frac{2C_{ij}}{|E_{ij}|}\right\}$$

where the minimum is taken over all points of the region and the boundary. Since $A > 0$, $C > 0$, $F \leq 0$ and all are bounded it follows that a positive minimum exists and for that minimum h

$$\beta_0 \geq \sum_{m=1}^{4} \beta_m,$$

thus suggesting diagonal dominance (Ames [7]). The field value U_{ij} is the weighted average of its four immediate neighbors and the forcing function G evaluated at (x_i,y_j). This relation is given pictorially by the well known five point star. Note that so far mn linear equations in the unknown field values have been assembled by discretizing the partial differential operator L.

Discretization of the Boundary Data

A very general boundary condition involves the specification of a linear combination of u and its normal derivative at each point of the boundary:

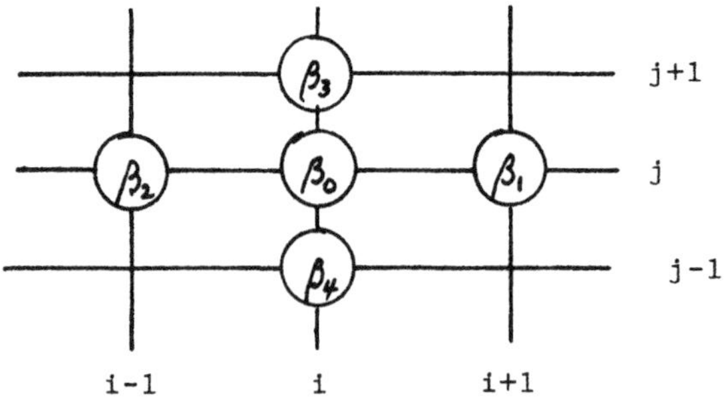

Figure 4. Five Point Computational Molecule.

$$\alpha u + \beta \partial u/\partial n = g \text{ on } \Gamma.$$

Second order accurate interior finite differences are used to approximate the normal derivative, which for our special problem is simply a partial derivative. The use of these interior finite difference approximations of the normal derivative term results in a matrix system that is larger than is absolutely necessary in the case of Dirichlet boundary conditions. This is the price paid for the ability to handle boundary conditions of the third kind.

The following forward/backward differences are used in the finite difference approximations to the boundary operators B. They are derived via the usual Taylor series manipulations.

$$u(x_{m+1},y_j) = (u(x_{m-1},y_j) - 4u(x_m,y_j) + 3u(x_{m+1},y_j))/2h + O(h^2)$$
$$u(x_0,y_j) = (-3u(x_0,y_j) + 4u(x_1,y_j) - u(x_2,y_j))/2h + O(h^2)$$
$$u(x_i,y_{n+1}) = (u(x_i,y_{n-1}) - 4u(x_i,y_n) + 3u(x_i,y_{n+1}))/2h + O(h^2)$$
$$u(x_i,y_0) = (-3u(x_i,y_0) + 4u(x_i,y_1) - u(x_i,y_2))/2h + O(h^2).$$

These second order accurate finite difference approximations, when inserted into the linear boundary operators, lead to four boundary molecules, as shown in Figure 5.

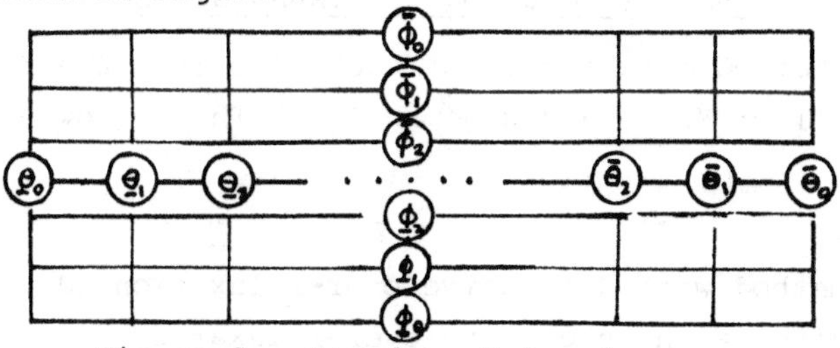

Figure 5. Boundary Molecules.

The following equations relate the boundary data functions to the symbols used in the boundary molecules of Figure 5.

Top: $\delta_{1,i}U_{i,n-1} - 4\delta_{1,i}U_{i,n} + [3\delta_{1,i} + 2h\varepsilon_{1,i}]U_{i,n+1} = 2h\lambda_{1,i}$

$\bar{\phi}_{2,i}U_{i,n-1} + \bar{\phi}_{1,i}U_{i,n} + \bar{\phi}_{0,i}U_{i,n+1} = \Lambda_{1,i}$

$1 \leqq i \leqq m$

Bottom: $[2h\varepsilon_{0,i} - 3\delta_{0,i}]U_{i,0} + 4\delta_{0,i}U_{i,1} - \delta_{0,i}U_{i,2} = 2h\lambda_{0,i}$

$\underline{\phi}_{0,i}U_{i,0} + \underline{\phi}_{1,i}U_{i,1} + \underline{\phi}_{2,i}U_{i,2} = \Lambda_{0,i}$

$1 \leqq i \leqq m$

Right: $\alpha_{1,j}U_{n-i,j} - 4\alpha_{1,j}U_{m,j} + [3\alpha_{1,j} + 2h\beta_{1,j}]U_{n+1,j} = 2h\gamma_{1j}$

$\bar{\theta}_{2,j}U_{m-1,j} + \bar{\theta}_{1,j}U_{m,j} + \bar{\theta}_{0,j}U_{m+1,j} = \Gamma_{1,j}$

$1 \leqq j \leqq n$

Left: $[2h\beta_{0,j} - 3\alpha_{0,j}]U_{0,j} + 4\alpha_{0,j}U_{1,j} - \alpha_{0,j}U_{2,j} = 2h\gamma_{0,j}$

$\underline{\theta}_{0,1}U_{0,j} + \underline{\theta}_{1,j}U_{1,j} + \underline{\theta}_{2,j}U_{2,j} = \Gamma_{0,j} \quad 1 \leqq j \leqq n$

STRUCTURE OF THE LINEAR SYSTEM

The linear system of the previous section is block tridiagonal as shown in Figure 6. Further each block, A_{ij} is an M by M tridiagonal matrix, except for the first and last rows. Consequently, direct solvers for block tridiagonal systems can be applied but that is not done here.

$$A = \begin{bmatrix} A_{11} & A_{12} & A_{13} & 0 & 0 & & \\ A_{21} & A_{22} & A_{23} & 0 & 0 & \emptyset & \\ 0 & A_{32} & A_{33} & A_{34} & 0 & & \\ & & & & A_{N-1,N-2} & A_{N-1,N-1} & A_{N-1,N} \\ & \emptyset & & & A_{N,N-2} & A_{N,N-1} & A_{NN} \end{bmatrix}$$

Figure 6. Block Structure of the Linear System.

ALGORITHMS FOR ITERATIVE SOLUTION OF LINEAR SYSTEMS

For a detailed description of each algorithm the reader is referred to Ames [7] or Young and Hageman [9]. The following methods were programmed.

Point Methods

1. Jacobi method with Successive over-relaxation (JOR)
2. Gauss-Seidel with Successive over-relaxation (SOR)
3. Symmetric Successive Overrelaxation (SSOR)

Block Methods

1. JOR by lines
2. SOR by lines
3. Alternating direction implicit method (ADI)

Descent Methods

1. Steepest descent
2. Conjugate gradient

Accelerations

1. Lyusternik method
2. Richardson method
   (in addition to SOR).

COMPUTER IMPLEMENTATION

All routines come in the four flavors as shown on the next page.

This discretization of the boundary operators generates 2m + 2n additional linear equations in the unknown (m+2)(n+2) field values -- a total of mn + 2(m+n) equations so far. Thus, only four equations are lacking to complete a square system. Note that the four corner field values $U_{0,0}$, $U_{m+1,0}$, $U_{0,n+1}$, and $U_{m+1,n+1}$ have not appeared in any of the equations so far. One might choose to make a corner grid value a weighted average of its neighbors. Our formulation sets all four corner points to the arbitrary value of zero. The particular choice of value does not matter. Thus, the final four equations of the (m+2)· (n+2) by (m+2)(n+2) linear system are

$$U_{0,0} = 0.0,$$

$$U_{m+1,0} = 0.0,$$

$$U_{0,n+1} = 0.0,$$

$$U_{m+1,n+1} = 0.0.$$

Indexing the Field Points

If the linear equations are arranged in a random fashion, then the corresponding coefficient matrix will exhibit little if any structure. If we label the unknown field points $U_{i,j}$, $0 \leqq i \leqq m+1$ and $0 \leqq j \leqq n+1$, in the natural row order (letting the row index i run first), then the corresponding coefficient matrix will have a banded structure that possesses distinct computational advantages. Note that other ordering are possible -- see Ames [7], Varga [8], or Young and Hageman [9].

An ordering is really just a bijective map from the doubly indexed approximate field values $U_{i,j}$ and the singly indexed components of the vector of unknowns in our linear algebraic equations. We will write our system of linear equations, arranged according to the natural row ordering as

$$AU = B.$$

The structure of the coefficient matrix A and the vector of inhomogeneities B is the subject of the next section. The map between the field values $U_{i,j}$, $0 \leqq i \leqq m+1$ and $0 \leqq j \leqq n+1$, and the vector components, labeled by U_k, $1 \leqq k \leqq MN$, where $M := m+2$ and $N := n+2$ is given by

Forward Map: $\quad k(i,j) = jm + i + 1$

Backward Map: $\quad i(k) = (k-1) \bmod M$

$\qquad\qquad\qquad\quad j(k) = (k-1) \operatorname{div} M.$

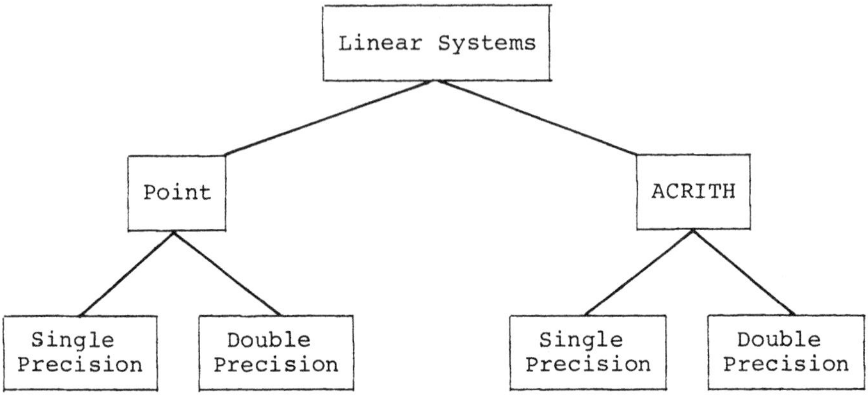

"Classical Method" "Interval Method"

Iteration Termination Criterion

The iteration is terminated when

$$\max_{i,j} |U_{i,j}^{(n+1)} - U_{i,j}^{(n)}| \leq \varepsilon$$

for the point method with a corresponding condition, using a maximum interval vector norm, when ACRITH is employed.

EXAMPLE PROBLEMS

All problems were run using the SOR Algorithm with the overrelaxation factor (ω) equal to one. All times are in microseconds and $0 < x < 1$, $0 < y < 1$,

I. A constant coefficient problem

$$u_{xx} + u_{yy} = 0, \quad u(x,0) = 0, \quad u(0,y) = 0,$$

$$u(x,1) = 1, \quad u(1,y) = 1$$

$$\varepsilon = 10^{-4}, \quad m = n = 7, \quad h = 0.125$$

	Point Solutions	Interval Solutions
System Generation Time	65968	66824
Iteration Time	3899148	69627276
Number of Iterations	49	49
Program Time	4,194,364	70,301,148

The point solutions failed to be in the guaranteed intervals at eight points. The maximum deviation is 5×10^{-6} from the interval boundary.

II. A linear source problem

$$-\frac{\partial}{\partial x}\left(p(x,y)\frac{\partial u}{\partial x}\right) - \frac{\partial}{\partial y}\left(p(x,y)\frac{\partial u}{\partial y}\right) + u = 0,$$

$p(x,y) = \exp(x^2 + y^2)$

$u(x,0) = 0, \quad u(0,y) = 0$

$u(x,1) = \frac{x}{1 - 0.9x}, \quad u(1,y) = \frac{y}{1 - 0.9y}$

$\varepsilon = 10^{-4}, \quad m = n = 7, \quad h = 0.125$

	Point Solutions	Interval Solutions
System Generation Time	628880	635988
Iteration Time	3263332	58407080
Number of Iterations	41	41
Program Time	4124612	59644352

The solution of this problem has very steep gradients near the left hand boundary ($x = 0$). The point solutions failed to lie in the guaranteed intervals at 20 points of the 81 grid points. The maximum deviation is 2×10^{-5} from the interval boundary.

III. Poisson equation with Dirichlet boundary conditions

$-(u_{xx} + u_{yy}) = 0.1/(x^2 + y^2), \quad 0 < x < 1, \quad 0 < y < 1$

$u(x,0) = u(x,1) = u(0,y) = u(1,y) = 0$

$\varepsilon = 10^{-4}, \quad m = n = 4, \quad h = 0.200$

	Point Solutions	Interval Solutions
System Generation Time	17404	17428
Iteration Time	264268	5633024
Number of Iterations	16	16
Program Time	374044	5888064

All point solutions lie within the guaranteed intervals. This equation has a singular point at the origin. The "sourcelike" solution is constructable by setting $u = f(\eta)$, $\eta = x^2 + y^2$ whereupon f satisfies the ordinary differential equation $(\eta f')' = -0.025\eta^{-1}$. The solution of that equation is $f(\eta) = A \ln \eta + B - 0.025(\ln \eta)^2$. Of course, the assumed data cannot be applied because of the singularity.

IV. Poisson equation with Dirichlet boundary conditions

$-(u_{xx} + u_{yy}) = \exp[-(x - \frac{1}{2})^2 - (y - \frac{1}{2})^2] \quad 0 < x < 1, \quad 0 < y < 1$

$u = 0$ on boundary

$\varepsilon = 10^{-4}, \quad m = n = 7, \quad h = 0.125$

	Point Solutions	Interval Solutions
System Generating Time	72612	70896
Iteration Time	3841036	71082140

	Point Solutions	Interval Solutions
Number of Iterations	48	49
Program Time	4144520	71754192

The point solutions <u>never</u> lie in the guaranteed intervals. The maximum deviation from the interval boundaries is 5×10^{-5}. Figures 7 and 7A illustrate this output.

V. <u>Exponential coefficients</u>

$$e^{y}u_{xx} + e^{x}u_{yy} = 0, \quad 0 < x < 1, \quad 0 < y < 1$$

$$u(x,0) = u(0,y) = 0$$

$$u(x,1) = u(1,y) = 1$$

$$\varepsilon = 10^{-4}, \quad m = n = 7, \quad h = 0.125$$

	Point Solutions	Interval Solutions
System Generation Time	83764	84072
Iteration Time	3902776	64125396
Number of Iterations	49	45
Program Time	4217176	64814000

The point solutions <u>never</u> lie in the guaranteed intervals with very bad agreement.

VI. <u>Negative exponential coefficients and variable boundary conditions</u>

$$e^{-y}u_{xx} + e^{-x}u_{yy} = 0, \quad 0 < x < 1, \quad 0 < y < 1$$

$$u(x,0) = u(0,y) = 0$$

$$u(x,1) = 1.0/(1 - 0.9x)$$

$$u(1,y) = 1.0/(1 - 0.9y)$$

$$\varepsilon = 10^{-4}, \quad m = n = 7, \quad h = 0.125$$

	Point Solutions	Interval Solutions
System Generation Time	83840	84096
Iteration Time	3584740	64131072
Number of Iterations	45	45
Program Time	3901672	64819336

Eleven of the point solutions failed to be in the guaranteed intervals. However, the deviation is not significant.

<u>DISCUSSION</u>

The ratio of the interval program time divided by the point program time is shown, for the individual cases, in the table below:

Figure 7

SYSTEM INDEX K	FIELD INDICES I	FIELD INDICES J	FIELD COORDINATES X(I)	FIELD COORDINATES Y(J)	POINT VALUES U(I,J)	INTERVAL VALUES UL(I,J)	INTERVAL VALUES UR(I,J)	
1	0	0	0.0000	0.0000	0.00000000E+00	0.0000000000E+00	0.0000000000E+00	
2	1	0	0.1250	0.0000	0.00000000E+00	0.0000000000E+00	0.0000000000E+00	
3	2	0	0.2500	0.0000	0.00000000E+00	0.0000000000E+00	0.0000000000E+00	
4	3	0	0.3750	0.0000	0.00000000E+00	0.0000000000E+00	0.0000000000E+00	
5	4	0	0.5000	0.0000	0.00000000E+00	0.0000000000E+00	0.0000000000E+00	
6	5	0	0.6250	0.0000	0.00000000E+00	0.0000000000E+00	0.0000000000E+00	
7	6	0	0.7500	0.0000	0.00000000E+00	0.0000000000E+00	0.0000000000E+00	
8	7	0	0.8750	0.0000	0.00000000E+00	0.0000000000E+00	0.0000000000E+00	
9	8	0	1.0000	0.0000	0.00000000E+00	0.0000000000E+00	0.0000000000E+00	
10	0	1	0.0000	0.1250	0.00000000E+00	0.0000000000E+00	0.0000000000E+00	
11	1	1	0.1250	0.1250	0.15150487E-01	0.151518E-01	0.151520E-01	*
12	2	1	0.2500	0.1250	0.24406508E-01	0.244088E-01	0.244091E-01	*
13	3	1	0.3750	0.1250	0.29514134E-01	0.295169E-01	0.295173E-01	*
14	4	1	0.5000	0.1250	0.31156331E-01	0.311591E-01	0.311595E-01	*
15	5	1	0.6250	0.1250	0.29516920E-01	0.295193E-01	0.295197E-01	*
16	6	1	0.7500	0.1250	0.24410784E-01	0.244125E-01	0.244128E-01	*
17	7	1	0.8750	0.1250	0.15153982E-01	0.151548E-01	0.151550E-01	*
18	8	1	1.0000	0.1250	0.00000000E+00	0.0000000000E+00	0.0000000000E+00	
19	0	2	0.0000	0.2500	0.00000000E+00	0.0000000000E+00	0.0000000000E+00	
20	1	2	0.1250	0.2500	0.24406508E-01	0.244088E-01	0.244091E-01	*
21	2	2	0.2500	0.2500	0.40216610E-01	0.402206E-01	0.402210E-01	*
22	3	2	0.3750	0.2500	0.49137831E-01	0.491426E-01	0.491432E-01	*
23	4	2	0.5000	0.2500	0.52027479E-01	0.520322E-01	0.520329E-01	*
24	5	2	0.6250	0.2500	0.49142614E-01	0.491467E-01	0.491473E-01	*
25	6	2	0.7500	0.2500	0.40223926E-01	0.402268E-01	0.402273E-01	*
26	7	2	0.8750	0.2500	0.24412468E-01	0.244139E-01	0.244142E-01	*
27	8	2	1.0000	0.2500	0.00000000E+00	0.0000000000E+00	0.0000000000E+00	
28	0	3	0.0000	0.3750	0.00000000E+00	0.0000000000E+00	0.0000000000E+00	
29	1	3	0.1250	0.3750	0.29514119E-01	0.295169E-01	0.295173E-01	*
30	2	3	0.2500	0.3750	0.49137831E-01	0.491426E-01	0.491432E-01	*
31	3	3	0.3750	0.3750	0.60354799E-01	0.603605E-01	0.603612E-01	*
32	4	3	0.5000	0.3750	0.64006388E-01	0.640121E-01	0.640129E-01	*
33	5	3	0.6250	0.3750	0.60360521E-01	0.603654E-01	0.603662E-01	*
34	6	3	0.7500	0.3750	0.49146652E-01	0.491501E-01	0.491507E-01	*
35	7	3	0.8750	0.3750	0.29521331E-01	0.295230E-01	0.295235E-01	*
36	8	3	1.0000	0.3750	0.00000000E+00	0.0000000000E+00	0.0000000000E+00	
37	0	4	0.0000	0.5000	0.00000000E+00	0.0000000000E+00	0.0000000000E+00	
38	1	4	0.1250	0.5000	0.31156316E-01	0.311591E-01	0.311595E-01	*
39	2	4	0.2500	0.5000	0.52027464E-01	0.520322E-01	0.520329E-01	*
40	3	4	0.3750	0.5000	0.64006329E-01	0.640121E-01	0.640129E-01	*
41	4	4	0.5000	0.5000	0.67912579E-01	0.679184E-01	0.679192E-01	*
42	5	4	0.6250	0.5000	0.64012110E-01	0.640170E-01	0.640178E-01	*
43	6	4	0.7500	0.5000	0.52036285E-01	0.520398E-01	0.520404E-01	*
44	7	4	0.8750	0.5000	0.31163514E-01	0.311652E-01	0.311657E-01	*
45	8	4	1.0000	0.5000	0.00000000E+00	0.0000000000E+00	0.0000000000E+00	

* - INTERVAL INCLUSION FAILED

1

Figure 7 continued

PAGE 2

SYSTEM INDEX K	FIELD INDICES I	J	FIELD COORDINATES X(I)	Y(J)	POINT VALUES U(I,J)	INTERVAL VALUES UL(I,J)	UR(I,J)	
46	0	5	0.0000	0.6250	0.00000000E+00	(0.0000000000E+00 ,	0.0000000000E+00)	
47	1	5	0.1250	0.6250	0.29516906E-01	(0.295193E-01 ,	0.295197E-01)	*
48	2	5	0.2500	0.6250	0.49142569E-01	(0.491467E-01 ,	0.491473E-01)	*
49	3	5	0.3750	0.6250	0.60360491E-01	(0.603654E-01 ,	0.603662E-01)	*
50	4	5	0.5000	0.6250	0.64012110E-01	(0.640170E-01 ,	0.640178E-01)	*
51	5	5	0.6250	0.6250	0.60365409E-01	(0.603696E-01 ,	0.603704E-01)	*
52	6	5	0.7500	0.6250	0.49150109E-01	(0.491531E-01 ,	0.491537E-01)	*
53	7	5	0.8750	0.6250	0.29523060E-01	(0.295245E-01 ,	0.295249E-01)	*
54	8	5	1.0000	0.6250	0.00000000E+00	(0.0000000000E+00 ,	0.0000000000E+00)	
55	0	6	0.0000	0.7500	0.00000000E+00	(0.0000000000E+00 ,	0.0000000000E+00)	
56	1	6	0.1250	0.7500	0.24410784E-01	(0.244125E-01 ,	0.244128E-01)	*
57	2	6	0.2500	0.7500	0.40223911E-01	(0.402268E-01 ,	0.402273E-01)	*
58	3	6	0.3750	0.7500	0.49146637E-01	(0.491501E-01 ,	0.491507E-01)	*
59	4	6	0.5000	0.7500	0.52036270E-01	(0.520398E-01 ,	0.520404E-01)	*
60	5	6	0.6250	0.7500	0.49150109E-01	(0.491531E-01 ,	0.491537E-01)	*
61	6	6	0.7500	0.7500	0.40229246E-01	(0.402313E-01 ,	0.402318E-01)	*
62	7	6	0.8750	0.7500	0.24415120E-01	(0.244161E-01 ,	0.244165E-01)	*
63	8	6	1.0000	0.7500	0.00000000E+00	(0.0000000000E+00 ,	0.0000000000E+00)	
64	0	7	0.0000	0.8750	0.00000000E+00	(0.0000000000E+00 ,	0.0000000000E+00)	
65	1	7	0.1250	0.8750	0.15153982E-01	(0.151548E-01 ,	0.151550E-01)	*
66	2	7	0.2500	0.8750	0.24412468E-01	(0.244139E-01 ,	0.244142E-01)	*
67	3	7	0.3750	0.8750	0.29521331E-01	(0.295230E-01 ,	0.295235E-01)	*
68	4	7	0.5000	0.8750	0.31163514E-01	(0.311652E-01 ,	0.311657E-01)	*
69	5	7	0.6250	0.8750	0.29523060E-01	(0.295245E-01 ,	0.295249E-01)	*
70	6	7	0.7500	0.8750	0.24415120E-01	(0.244161E-01 ,	0.244165E-01)	*
71	7	7	0.8750	0.8750	0.15156150E-01	(0.151566E-01 ,	0.151569E-01)	*
72	8	7	1.0000	0.8750	0.00000000E+00	(0.0000000000E+00 ,	0.0000000000E+00)	
73	0	8	0.0000	1.0000	0.00000000E+00	(0.0000000000E+00 ,	0.0000000000E+00)	
74	1	8	0.1250	1.0000	0.00000000E+00	(0.0000000000E+00 ,	0.0000000000E+00)	
75	2	8	0.2500	1.0000	0.00000000E+00	(0.0000000000E+00 ,	0.0000000000E+00)	
76	3	8	0.3750	1.0000	0.00000000E+00	(0.0000000000E+00 ,	0.0000000000E+00)	
77	4	8	0.5000	1.0000	0.00000000E+00	(0.0000000000E+00 ,	0.0000000000E+00)	
78	5	8	0.6250	1.0000	0.00000000E+00	(0.0000000000E+00 ,	0.0000000000E+00)	
79	6	8	0.7500	1.0000	0.00000000E+00	(0.0000000000E+00 ,	0.0000000000E+00)	
80	7	8	0.8750	1.0000	0.00000000E+00	(0.0000000000E+00 ,	0.0000000000E+00)	
81	8	8	1.0000	1.0000	0.00000000E+00	(0.0000000000E+00 ,	0.0000000000E+00)	

1

NUMBER OF VERTICAL MESH LINES: 9
NUMBER OF HORIZONTAL MESH LINES: 9
UNIFORM MESH SPACING: 0.125000

SPELL RUN STATISTICS

SPELL RUN START: 21:34:58 ON 12/23/84
SPELL RUN END: 21:35:04 ON 12/23/84

NUMBER OF ITERATIONS: 48
MAXIMUM NUMBER OF ITERATIONS: 500
RELATIVE TERMINATION CRITERION: 0.10000000E-03
RELATIVE DIFFERENCE OF LAST ITERATES: 0.99615223E-04
MAX-NORM OF THE RESIDUAL VECTOR: 0.13448298E-04
SYSTEM GENERATION TIME (MICROSECS): 72612
ITERATION TIME (MICROSECS): 3841036
PROGRAM TIME (MICROSECS): 4144520

	I	II	III	IV	V	VI
$\frac{\text{Interval Time}}{\text{Point Time}}$	16.8	14.5	15.7	17.3	15.4	16.6

To obtain guaranteed bounds on the solution one must pay a price -- an increase in the computation time of 15-17 times when the dot product of ACRITH is not used. If the micro-coded dot product is used this factor will be less than 2.

REFERENCES

1. Kulisch, U. W. and Miranker, W. L., Computer Arithmetic in Theory and Practice, Academic Press, New York, 1981.

2. Kulisch, U. W. and Miranker, W. L., eds., A New Approach to Scientific Computation, Academic Press, New York, 1983.

3. IBM, ACRITH High-Accuracy Arithmetic Subroutine Library, General Information Manual, IBM Publication Number GC33-6163.

4. IBM, ACRITH High-Accuracy Arithmetic Subroutine Library, Program Description and User's Guide, First Edition (October 1983), IBM Publication Number SC33-6164-0.

5. Samarski, A. A. and Tychonov, A. N., Partial Differential Equations of Mathematical Physics, Volumes I and II, Holden-Day, San Francisco, 1967.

6. Ames, W. F., Nonlinear Partial Differential Equations in Engineering, Academic Press, New York, I 1965, II 1972.

7. Ames, W. F., Numerical Methods for Partial Differential Equations, Academic Press, New York, 1977.

8. Varga, R. S., Matrix Iterative Analysis, Prentice Hall, 1962.

9. Young, P. M. and Hagemen, L. A., Applied Iterative Methods, Academic Press, 1981.

10. Moore, R. E., Interval Analysis, Prentice-Hall, New Jersey, 1966.

Case Studies for Augmented Floating-Point Arithmetic

W.L. Miranker, M. Mascagni[1]
IBM T.J. Watson Research Center
Yorktown Heights, New York

S. Rump
IBM Development Laboratory
Boeblingen, W. Germany

1. Introduction

Much of scientific computation involves operations with higher data types such as vectors, matrices, complex quantities, and intervals of these. A well implemented conventional computer arithmetic gives good results with floating-point numbers taken two at time. However, using this arithmetic to implement computations with the higher data types can return arbitrarily bad results. Kulisch and Miranker [4] have proposed a computer arithmetic that furnishes an optimal accuracy for the dyadic arithmetic operations for all of these higher data types. Its implementation is based on the introduction of an accurate inner product to the standard floating-point arithmetic operation set.

Let the rounding in a computer be represented by a rounding operator (denoted $\square$) that maps the real numbers into the computer representable (i.e., floating-point) numbers S; $\square : \mathbb{R} \rightarrow S$. Well designed computers should have a floating-point arithmetic that delivers $a \boxdot b = \square\,(a \circ b)$, for $a,\ b \in\ S$ and for $\circ \in \{\ +\ ,\ -\ ,\ \times,\ /\ \}$. Here $\boxdot$ denotes to the computer implementation of the operation $\circ$. For spaces of vectors $\mathbb{R}^n$ or matrices $\mathbb{R}^{m \times n}$, the rounding $\square$ is defined componentwise. Then with the accurate inner product (denoted $\boxdot$), or the fifth operation as we shall also call it, we can achieve $a \boxdot b = \square\,(a \circ b)$ for $a,\ b\ \in\ S,\ S^n,\ S^{m \times n}$ with $\circ$ being any admissible dyadic computation in those spaces, or between those spaces. This definition of the operations of computer arithmetic is described by the term semimorphism. Similar as-

[1] Under the auspices of a student interaction agreement with the Courant Institute

sertions are available for the interval data types, although we shall not deal further with those types here. Between the exact unrounded result of an operation and its value furnished by the semimorphic definition there can be found no other computer representable data type. Thus such operations are optimal. For examples and details concerning these assertions, see [4,5,6] and the article of U. Kulisch in these proceedings. An early class of applications of this new computer arithmetic consists of the so-called E-method. This method [2,3,12] uses the accurate inner product in concert with directed roundings, interval arithmetic, and residual correction to produce computations of both high accuracy and guarantees, i.e., to produce validated computation. The guarantees include tight bounds for the solution of the actual problem, bounds often providing least significant bit accuracy.

Although supplying the numerical analyst with a more exacting tool for probing, E-methods are time consuming and have as yet only enjoyed only moderate popularity among methods employed for day-to-day scientific computations. The intrinsic amount of computation required for E-methods also makes them somewhat impractical production nature computation. A related question is: can the accurate inner product be routinely employed in standard numerical algorithms with some benefit? Here we shall describe numerical experiments that answer this question in the affirmative for some algorithms and give some seemingly negative results for others.

Two types of improvements have been observed. The first is that an iterative computation may require fewer iterations when implemented with the accurate inner product for the same heuristic convergence criterion as compared to the conventional implementation. In addition, for a fixed number of iterations, use of the accurate inner product may return results of higher quality than the corresponding conventional four operation computation. The computations that show positive results are for 1. ill-posed problems as exemplified by harmonic continuation, 2. stiff differential equations, 3. the QR algorithm for determining the eigenelements of a matrix, 4. the conjugate gradient algorithm for solving linear systems. In the cases 1 and 2, the iteration which delivers the favorable results is the method of residual correction applied to refining the numerical solution of a linear system arising in the numerical method. These four studies will be described in detail successively in sections 2-5. Some conclusions are offered in section 6. For additional examples and further details see [8, 10] , The numerical evidence makes a strong case for augmenting the standard floating-point arithmetic operations with the accurate inner product.

We shall use the following notation where single precision means 7 decimal digits while double precision means 17 decimal digits.

S_4 : computation done in Single precision with the 4 operations: ⊞, ⊟, ⊠, ◫.

S_5 : computation done in Single precision with the 5 operations: ⊞, ⊟, ⊠, ◫, ⊡.

D_4 : computation done in Double precision with 4 operations.

D_5 : computation done in Double precision with 5 operations.

2. Ill-posed problems

A Cauchy problem for Laplace's equation is a standard example of an ill-posed problem:

$$\frac{\partial^2 u}{\partial x^2} + \frac{\partial^2 u}{\partial y^2} = 0, \quad |x| < \infty,\ y > 0,$$

$$\frac{\partial u}{\partial y}(x, 0) = \frac{1}{n} \sin\ nx.$$

The solution of this problem is

$$u(x, y) = \frac{1}{n^2} \sin\ nx\ \sinh\ ny.$$

Thus as $n \to \infty$, the initial data, $\frac{1}{n}\sin\ nx \to 0$, whereas for $y > 0$, the solution tends to infinity. Then arbitrarily small perturbations of initial data may lead to arbitrarily large changes in the solution. That is, the problem solution does not depend continuously on its initial data.

An ill-posed problem is specified in the following definition.

Definition: A problem is well-posed of its solution exists, is unique and depends continuously on its data. A problem is ill-posed if it is not well-posed.

Many ill-posed problems take the form of a Fredholm integral equation of the first kind

$$\mathcal{K} f := \int_a^b K(x, y)\, f(y)\, dy = g(x), \quad x \in [a, b], \tag{2.1}$$

and the Riemann-Lebesque lemma shows why such a problem is in general ill-posed:

Riemann-Lebesque Lemma: Let $K(x, y) \in C_y[a, b]$. Then

$$\lim_{n \to \infty} \int_a^b K(x, y) \sin\ ny\, dy = 0.$$

Now let the perturbed data $g + \Delta g$ correspond to the solution $f + \Delta f$, and let $\Delta f = \sin\ ny$. Then

$$\Delta g = \int_a^b K(x, y) \sin\ ny\, dy.$$

Then by the Riemann-Lebesque lemma, arbitrarily small changes Δg (obtained by increasing n here) in the data of the integral equation (2.1) correspond to $O(1)$ changes Δf in its solution.

Upon compatibly discretizing the integral equation, we obtain

$$[K] f = g. \tag{2.2}$$

Here f and g are N-vectors and $[K]$ is an N×N matrix.

This discretized version of the ill-posed integral equation is itself an ill-conditioned linear system. To see this, suppose that the kernel $K(x,y)$ is symmetric and positive definite. Then the spectrum of K consists of a sequence of eigenvalues (of finite multiplicity) tending to zero.

$$\lambda_1 \geq \lambda_2 \geq \lambda_3 \geq \ldots > 0.$$

Moreover compatibility of the discretization causes the spectrum of $[K]$ to converge to the spectrum of $\mathcal{K}$ as N $\to \infty$. Let

$$\mu_1 \geq \mu_2 \geq \ldots \geq \mu_N$$

be the spectrum of $[K]$. Then since $\mu_1 \to \lambda_1$ and $\mu_N \to \lambda_N$, we obtain

$$\frac{\mu_1}{\mu_N} \to \frac{\lambda_1}{\lambda_N} \to \infty,$$

as $N \to \infty$. That is, the condition number of K,

$$cond\,[K] \to \infty \quad \text{as} \quad N \to \infty. \tag{2.3}$$

Examples of classical ill-posed problems are: the backward heat equation, analytic (harmonic) continuation, the Cauchy problem for Laplace's equation, and numerical differentiation.

Tradition had it that ill-posed problems were mathematical pathologies with no relevance in applications [1]. A current view is quite oppositely directed as ill-posed problems arise in such contemporary fields as inverse scattering, tomography, x-ray crystallography and Fourier optics.

For instance, the basic problem of Fourier optics is the solve the integral equation

$$\int_{-\frac{x}{2}}^{\frac{-x}{2}} S(x-y)\, f(y)\, dy = g(x).$$

Here

$S(x)$, the point spread function, is a function whose Fourier transform has compact support, say in the interval $[-\Omega, \Omega]$. For instance, $S(x) = (\sin \pi\Omega x)/(\pi x)$,

$f(x)$ is the wave function in the object plane, i.e., it corresponds to the object itself,

$g(x)$ is the wave function in the image plane, i.e., the noiseless image.

Harmonic continuation: Our numerical experiment deals with harmonic continuation. Let f be a harmonic function in a disc of radius R centered at the origin, and let $r < R$. Then the Poisson integral formula expresses the continuation of f from the circle $Re^{i\theta}$ to the circle $Re^{i\theta}$, $\theta \in [0, 2\pi)$:

$$f(re^{i\theta}) = \frac{1}{2\pi} \int_0^{2\pi} f(Re^{i\alpha}) \frac{R^2 - r^2}{R^2 - 2rR\cos(\theta - \alpha) + r^2}\, d\alpha.$$

We discretize this problem by using Simpson's rule and N equally spaced points on the circle $re^{i\theta}$. Let the N-vector f_R resp. f_r denote the discretization of $f(Re^{i\theta})$ resp. $f(re^{i\theta})$.
The resulting linear system

$$[P]\, f_R = f_r \tag{2.4}$$

is solved in three ways:

1. Gaussian elimination in single precision with 4 operations (S_4).

2. The same as 1 with double precision residual correction. This mixed mode of computation is denoted M.

3. The same as 2, the correction done in single precision but with 5 operations (S_5).

Results of these computations are displayed in Figures 2.1-2.6 where norms of relative errors $\|e\|$ are plotted. Unless otherwise indicated, the maximum norm (on the N points of discretization) is used.

Figure 2.1 contains a plot of the relative algebraic errors (relative to the exact solutions of the discretized linear system). The errors increase with decreasing r as expected, since this represents an increase in the continuation stepsize. The values of the curve corresponding to S_5 are of the order of the number of digits in the single precision floating point used in the computer.

Figures 2.2-2.6 contain plots of the relative errors (relative to the exact values of $f(Re^{i\theta})$).

In Figures 2.2-2.4, we see that the results for M and for S_5 are essentially the same for small continuation stepsize (i.e., for r near unity). As r decreases, M fails catastrophically producing extremely large errors, whereas S_5 continues with a gradual degradation. Increasing $N(200 \rightarrow 300 \rightarrow 400)$, increases the ill-conditioning of the linear system (cf. (2.3)), resulting in an earlier onset of the catastrophic failure in question and a worsening of the gradual degradation of S_5. For small values of r, S_5 yields unacceptable errors, albeit noncatastrophic ones. For moderate r, S_5 yields good results which are moreover superior to those delivered by M.

In Figures 2.5 and 2.6, we plot errors for a process of repeated continuation. That is, we start with $f(re^{i\theta})$ and continue it to produce $f((r+\Delta r)e^{i\theta})$, $f((r+2\Delta r)e^{i\theta})$, ..., each by continuation from its predecessor. Results are given for $r = .7$, $N = 50$ and 100 and for various values of Δr. Each different choice of Δr corresponds to a different plot: a pair of curves corresponding to repeated continuation using M and S_5 for solving the corresponding discretized linear systems of the form (2.4).

In all cases, the first step of continuation resulted in similar errors. For subsequent continuation steps, the two methods produce differing results with S_5 always being superior to M. However, in many cases the errors furnished by both methods are not acceptable.

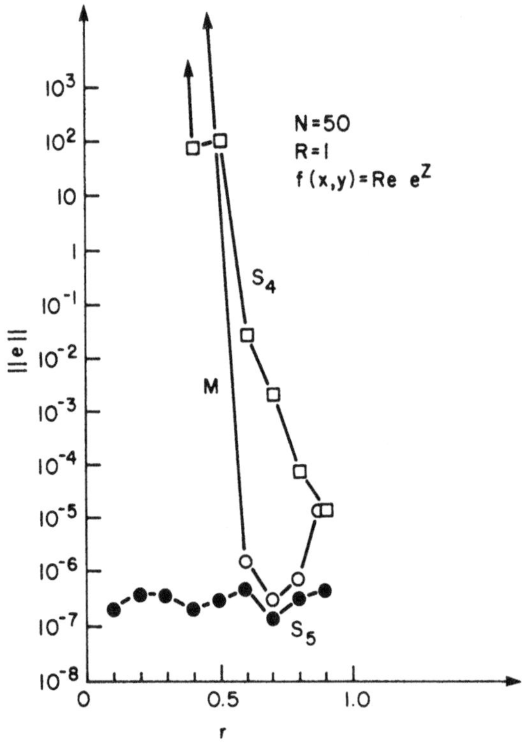

Fig. 2.1: Relative Algebraic Errors

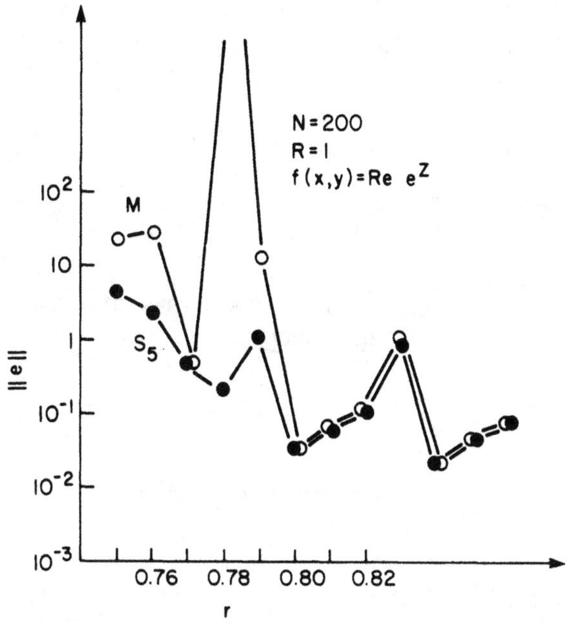

Fig. 2.2: Relative Errors, $N = 200$

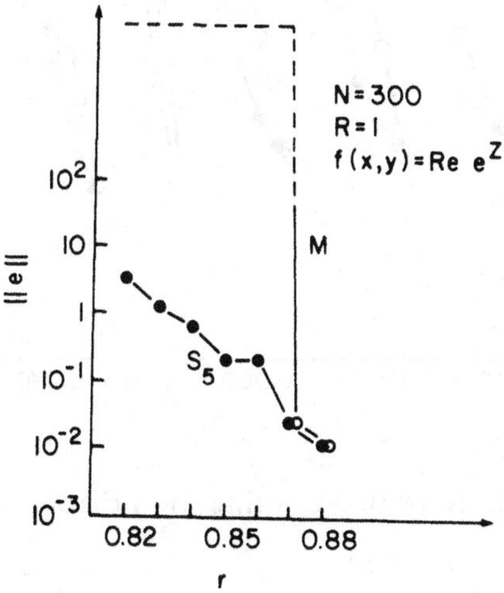

Fig. 2.3: Relative Errors, $N = 300$

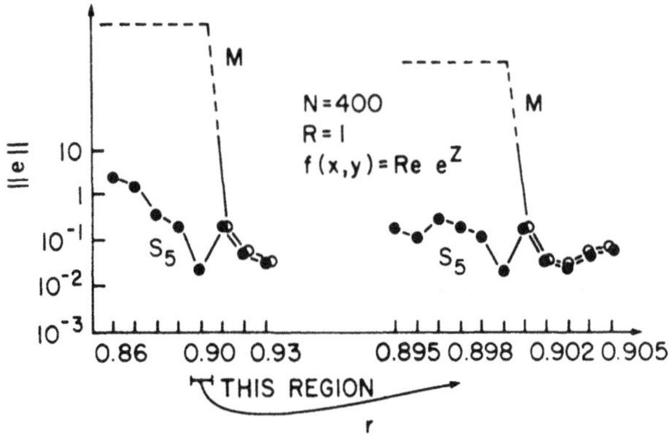

Fig. 2.4: Relative Errors, $N = 400$

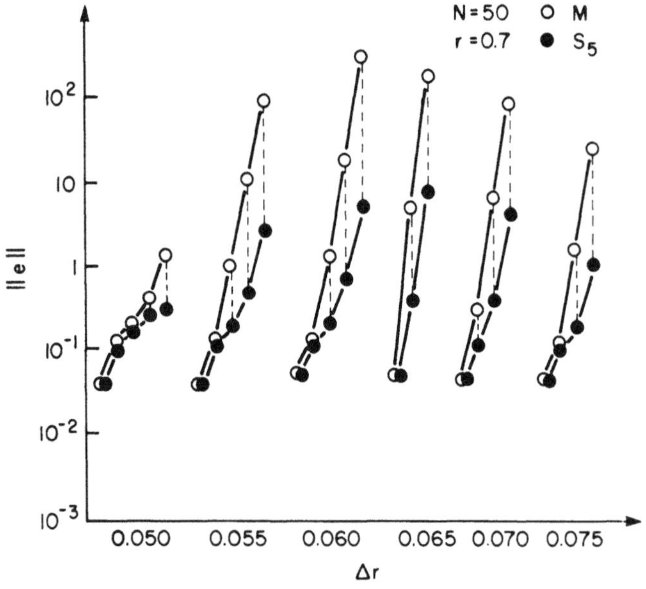

Fig. 2.5: Repeated Continuation Errors, $N = 50$

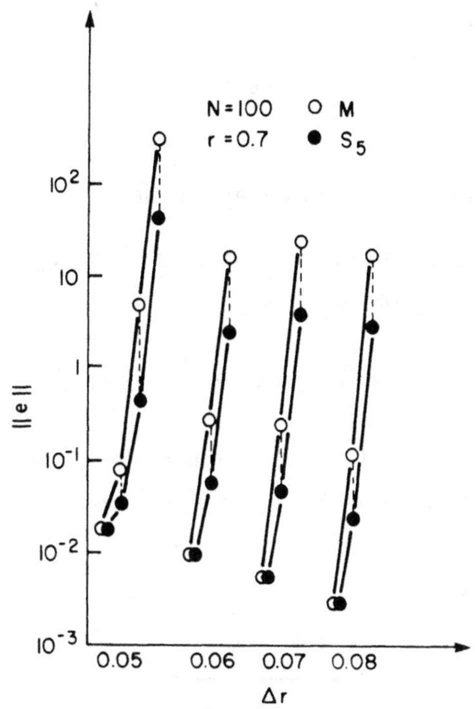

Fig. 2.6: Repeated Continuation Errors, $N = 100$

3. Stiff differential equations

The numerical treatment of a stiff differential equation [9] usually results in an ill-conditioned problem. To see this consider the initial value problem for an N-th order system

$$\begin{aligned} \frac{dy}{dt} &= -Ay, \quad t \in [0, 1], \\ y(0) &= y_0, \end{aligned} \tag{3.1}$$

of ordinary differential equations. Let the eigenvalues λ_i, $i = 1, ..., N$ of A have the property

$$0 < \lambda_1 < \lambda_2 < ... < \lambda_N,$$

where moreover, the ratio

$$\frac{\lambda_N}{\lambda_1} \tag{3.2}$$

is large (values of 10^{20} to 10^{30} are known in applications).

Now discretize (3.1) using the trapezoidal formula. The result is the following so-called A-stable difference approximation to (3.1).

$$(I - \frac{1}{2} hA)\, Y_n = (I + \frac{1}{2} hA)\, Y_{n-1}.$$

Here h is the mesh increment, and Y_n is the value of the numerical approximation to $y(nh)$, $n = 0, 1, \ldots$. Then to find Y_n, given Y_{n-1}, requires solving a linear system whose condition number

$$\frac{1 + \frac{1}{2} h\lambda_N}{1 - \frac{1}{2} h\lambda_1}$$

is large with the value of $h\lambda_N/\lambda_1$. The latter quantity is one measure of the stiffness of the differential equation in (3.1).

The numerical treatment deals with two nonlinear stiff systems. These systems are discretized using the trapezoidal formula. Then to find Y_n given Y_{n-1} requires solving a system of nonlinear equations. This nonlinear system is solved by using Newton's method with the value of the initial iterate taken as Y_{n-1} itself. Each iteration within Newton's method requires the solution of a linear system (the latter being the more ill-conditioned with the increasing stiffness of the original system of differential equations). This linear system is solved by the two methods previously used in section 2, namely M and S_5.

Iteration within Newton's method is continued until the typical numerical convergence criterion (invariance of successive iterates to some tolerance) is met. Call this number of iterations $N_n(\varepsilon)$, where ε is the tolerance. Since we do not know the exact solutions of the equations, our experiments are constructed as follows. We solve each system using a single step of mesh width, $h = 10^{-4}$. Then we repeat the procedure using ten steps of the mesh width $h = 0^{-5}$. The results now follow.

1) Stiff Differential System from Enzyme Kinetics [7].

$$y'_1 = 10^{11}(-3y_1y_2 + .0012y_4 - 9y_1y_3)$$

$$y'_2 = -3 \times 10^{11} y_1y_2 + 2 \times 10^7 y_4$$

$$y'_3 = 10^{11} \times (-9y_1y_3 + .001y_4)$$

$$y'_4 = 3y_1y_2 - .0012y_4 + 9y_1y_3$$

$$y_1(0) = 3.365 \times 10^{-7};\ y_2(0) = 8.261 \times 10^{-3}$$

$$y_3(0) = 1.642 \times 10^{-3};\ y_4(0) = 9.38 \times 10^{-6}$$

		h	$y_1(10^{-4})$	$y_2(10^{-4})$	$y_3(10^{-4})$	$y_4(10^{-4})$
M	10 steps	10^{-5}	2.02×10^{-7}	3.72×10^{-3}	6.19×10^{-3}	9.52×10^{-6}
	1 step	10^{-4}	7.30×10^{-8}	-6.88×10^{-3}	1.68×10^{-2}	9.64×10^{-6}
S_5	10 steps	10^{-5}	2.85×10^{-7}	3.72×10^{-3}	6.19×10^{-3}	9.43×10^{-6}
	1 step	10^{-4}	2.50×10^{-7}	4.01×10^{-3}	5.89×10^{-3}	9.69×10^{-6}

2) Stiff Differential System from Chemical Kinetics [7].

$$y'_1 = y_3 - 100 y_1 y_2$$

$$y'_2 = y_3 + 2 y_4 - 100 y_1 y_2 - 2 \times 10^4 y_2^2$$

$$y'_3 = y_3 + 100 y_1 y_2$$

$$y'_4 = y_4 + 10^4 \times y_2^2$$

$$y_1(0) = y_2(0) = 1;\ y_3(0) = y_4(0) = 10^{-8}$$

		h	$y_1(10^{-4})$	$y_2(10^{-4})$	$y_3(10^{-4})$	$y_4(10^{-4})$
M	10 steps	10^{-5}	0.99	0.33	0.0054	0.33
	1 step	10^{-4}	0.99	−0.0049	0.0049	0.33
S_5	10 steps	10^{-5}	0.99	0.33	0.0054	0.33
	1 step	10^{-4}	0.99	0.33	0.0066	0.33

When M is employed, $N_n(\varepsilon)$ varies between 3 and 8. For every case tried, $N_n(\varepsilon) = 2$ when the method S_5 is employed, a considerable savings. The iterations for method for S_5 were carried out in interval arithmetic in an E-method environment using ACRITH (see the article by S. Rump in these proceedings). In fact, we found that in this mode $N_n(\varepsilon) = 1$, since after one iteration, ACRITH supplied a guarantee of convergence moreover to within the required tolerance. If we ignore the ACRITH guarantee, an additional iteration would be required to confirm convergence according to the conventional tolerance test of the experiment at hand. Thus the value $N_n(\varepsilon) = 2$ is, in fact, an estimate.

The results for Y produced by S_5 for $h = 10^{-4}$ are nearly reconfirmed by the recalculation for $h = 10^{-5}$. The corresponding results for M are quite far from agreement which creates the need of at least another computational pass, say with $h = 10^{-6}$. This is a massive additional cost for M compared to S_5 .

4. QR Algorithm

The QR algorithm is a standard method for computing the eigenvalues $\lambda_1, \ldots, \lambda_N$ of an $N \times N$ matrix [11]. It is based on the so-called QR factorization of a matrix. Any real $N \times N$ matrix can be written as the product of an orthogonal matrix and an upper triangular matrix. This decomposition is in fact unique with the added constraint that the orthogonal matrix have a row whose elements have specified signs. The QR factorization can be viewed as the orthonormalization of the columns of A (say via the Gram-Schmidt procedure). The Q matrix is the resulting matrix of orthonormalized column vectors. The R matrix is necessarily upper triangular as the j-th column vector in A is at most a linear combination of the first j column vectors in Q.

The given matrix A_0 is factored into $Q_0\, R_0$. A new matrix A_1 is computed as $R_0\, Q_0$. Then A_1 is factored as $Q_1\, R_1$, etc. Note that all the matrices A_i, $i = 0, 1, 2, \ldots$ are similar (and therefore isospectral), since $A_0 = Q_0\, R_0 \Rightarrow R_0 = Q_0^T A_0$, so $A_1 = R_0\, Q_0 = Q_0^T A_0 Q_0$, etc. Indeed the A_i are unitarily equivalent. A_n converges to $\Lambda = diag(\lambda_N, \lambda_{N-1}, \ldots, \lambda_1)$ provided the eigenvalues of A_0 are distinct in absolute value. The QR algorithm also provides a method for computing the eigenvectors of A_0. Indeed

$$A_n = \left(\prod_{i=0}^{n-1} Q_i\right)^T A_0 \prod_{i=0}^{n-1} Q_i ,$$

so that

$$\prod_{i=0}^{n-1} Q_i \, A_n \left(\prod_{i=0}^{n-1} Q_i\right)^T = A_0.$$

Since A_n converges to a diagonal matrix, $\prod_{i=0}^{n-1} Q_i$ must converge to the matrix whose columns are the eigenvectors of A_0.

If A_0 has a band structure, then the A_n will have the same band structure. For this reason real symmetric matrices are reduced to tridiagonal form by the method of Householder prior to applying the QR algorithm. This results in a reduction of computational and storage requirements. In the case of non-symmetric matrices the method of Householder is used to reduce the matrix to upper Hessenberg form to save space and time in the execution of the QR algorithm. The QR factorization of a matrix is carried out by either direct orthonormalization of the columns of the given matrix or by carrying out this procedure indirectly with the method of Householder. (We found that Gram-Schmidt gave superior computational performance in all cases.)

A stopping rule for the computation is based on the property that A_n converges to a diagonal matrix when A_0 is symmetric. Thus the sum of the squares of the diagonal elements of A_n must converge to the sum of the squares of the eigenvalues of A_0. The latter sum is just the sum of the squares of all of the elements of A_0. Let $A_0 = \left(A_0(i, j)\right)$, $A_n = \left(A_n(i, j)\right)$. Then the stopping rule is

$$\left| \sum_{i,j=1}^{N} A_0^2(i, j) - \sum_{i=1}^{N} A_n^2(i, i) \right| < \varepsilon \sum_{i,j=1}^{N} A_0^2(i, j).$$

These observations come from properties of the Frobenius norm (denoted $\|\cdot\|_F$) of a matrix. The Frobenius norm of a matrix is the L^2 norm of the $m\times n$ matrix taken as a vector in $\mathbb{R}^{mn}$. Multiplication of a vector by an orthogonal matrix leaves that vectors norm unchanged. Thus

$$\|A\|_F = \|AQ\|_F = \|QA\|_F,$$

where $QQ^T = I$.

Since

$$A_0 = \prod_{i=0}^{n-1} Q A_n \left(\prod_{i=0}^{n-1} Q\right)^T$$

$$\| A_0 \|_F = \| \prod_{i=0}^{n-1} Q_i \, A_n \left(\prod_{i=0}^{n-1} Q\right)^T \|_F = \| A_n \|_F \underset{n \to \infty}{\to} \| \Lambda \|_F.$$

There are, of course, other stopping rules.

Using Gram-Schmidt orthonormalization in the QR factorization requires a significant number of inner products to be computed each iteration. To illustrate this, let $A_0 = \text{col}\,(a_1, \dots, a_N)$, $Q = \text{col}\,(q_1, \dots, q_N)$ and the elements of $R = (r_{ij})$. Then the QR factorization is given by:

$$\begin{aligned} &1. \quad \gamma_i = a_i - \sum_{k=1}^{i-1} (q_k, a_i)\, q_k, \quad i = 1, 2, \dots, N, \\ &2. \quad r_{ki} = \begin{cases} (q_k, a_i), & k \neq i, \\ (\gamma_i, \gamma_i)^{1/2}, & k = i, \end{cases} \\ &3. \quad q_i = \frac{\gamma_i}{r_{ii}}, \quad i = 1, 2, \dots, N. \end{aligned} \tag{4.1}$$

Each r_{ij} is an inner product, and the computation of γ_i componentwise can be viewed as the inner product of vectors of length i or each of the N components of γ_i.

For computing the convergence criteria, the original sum of the squares of all the elements of A_0 is an inner product, as is the repeated computation of the sum of the squares of the diagonal elements of A_n.

The QR algorithm was studied comparatively using the four computational modes S_4, S_5, D_4, D_5. The matrix used in these computations arises in finite-difference applications. It is tridiagonal with 2 on the main diagonal and -1 on the off-diagonals. The spectrum of this matrix is known in closed form. We plot the number of iterations required to satisfy the convergence criterion versus ε, as well as the square root of the sum of the squares of the difference between the computed and the exact eigenvalues versus ε. Calculations were limited to 1001 iterations in the single precision experiments and to 5001 iterations in the double precision experiments.

With $N = 5$ the S_4 (computations) fail to converge for $\varepsilon = 10^{-5}$, a relatively large value (see Fig. 4.1). (Figures are found at the end of the section.) S_5 levels off to an iteration number (approximately 31) that satisfies all the successively more demanding epsilons. At first S_5 performs comparably to D_4 and D_5, but for values of ε less than 10^{-7}, S_5 also levels off. The L^2 error vs. the number of iterations plots (see Fig. 4.2) show that S_5 achieves the same reduction of error per iteration as does D_4 and D_5 while S_4 is considerably less efficient. With this small value of N, D_4 and D_5 behave exactly the same throughout the experimental range. Fig. 4.3 shows the accuracy of the computed results as a function of ε.

With $N = 20$ (see Figs. 4.4, 4.5, 4.6), the behavior of S_4 and S_5 relative to each other and relative to D_4 and D_5 remains unchanged. However, the difference in behavior seen between S_4 and S_5 is now seen between D_4 and D_5, the latter at very small values of epsilon. As in the single precision case, D_4 fails to satisfy the stopping criterion sooner than D_5 does. Notice that when both D_4 and D_5 perform the limiting number 5001 iterations, D_5 produces a result two orders of magnitude better than D_4. This behavior is replicated in the case $N = 60$ where the differences between D_4 and D_5 are even more clearly seen (see Figs. 4.7, 4.8, 4.9). In Figs. 4.1, 4.3, 4.4, 4.6, 4.7, and 4.9, the plots for S_5 have horizontal segments which extend below $\varepsilon = 10^{-7}$ (the single precision mantissa having 7 digits, approximately). This is an artifact of the form of the convergence criteria in the S_5 regime; two inner products are computed and compared, and in S_5, they always agree. A similar effect is seen in Fig. 4.9 for D_5 and $\varepsilon \leq 10^{-15}$ (corresponding to the double precision mantissa length of 17 digits).

Note that computations are sometimes performed in inpractical cases which would never be used in a real production setting. This is done only for purposes of our study.

Conclusions

1. Five operations allow convergence criteria to be evaluated more accurately avoiding unnecessary extra computation.
2. The individual QR factorizations are done more accurately with five operations resulting in a greater efficiency per iteration.
3. The phenomena described in 1 and 2 here are present in both single and double precision, the only significant difference being the location in which four and five operations results diverge from each other.

To better understand these results, we determine the number of roundings occurring in the computation of a single QR decomposition of an $N \times N$ matrix with four and five operations. The savings in roundings seen with five operations comes from (i) the explicit inner products ($2N-1$ roundings), and (ii) replacing the summation in (4.1)1 required to compute the components of γ_i with one inner product of length i. The total number of roundings for the QR factorization of an $N \times N$ matrix are:

4 operations	5 operations
$4N^3 - 2N^2 - N$	$\frac{7}{2}N^2 - \frac{N}{2}$

(See [8, appendix 1] for details.)

This difference between $O(N^3)$ for four operations and $O(N^2)$ for five operations is a significant factor in explaining our observations. Each rounding operation introduces a certain amount of noise into the computation. With a given noise level only so much information can be extracted from a computation.

Non-convergence can also be explained by noise. As the iteration progresses, the computed A_{n+1} is no longer isospectral to A_0. Each iteration brings about a small perturbation in the spectrum of A_{n+1} relative to A_n. This in turn makes convergence that relies on spectral information computed from A_0 increasingly harder to obtain. To gain some feeling for this effect, we perform a forward roundoff error analysis in the case $N = 2$. Let $\Box\, a = a + aO(\varepsilon)$. Here ε is the machine epsilon. If terms of $O(\varepsilon^2)$ are neglected, the resulting q_{ij} and r_{ij} are:

4 operations	5 operations
$r_{11} + \frac{5}{2}O(\varepsilon)r_{11} + \cdots$	$r_{11} + \frac{3}{2}O(\varepsilon)r_{11} + \cdots$
$q_{11} + \frac{7}{2}O(\varepsilon)q_{11} + \cdots$	$q_{11} + \frac{5}{2}O(\varepsilon)q_{11} + \cdots$
$q_{21} + \frac{7}{2}O(\varepsilon)q_{21} + \cdots$	$q_{11} + \frac{5}{2}O(\varepsilon)q_{21} + \cdots$
$r_{12} + \frac{13}{2}O(\varepsilon)r_{12} + \cdots$	$r_{12} + \frac{7}{2}O(\varepsilon)r_{12} + \cdots$

$$\text{4 ops: } q_{12} + 4O(\varepsilon)q_{12} + 11O(\varepsilon)\left[\frac{r_{12}q_{11}}{\gamma_{12}} + \frac{r_{12}q_{11}\gamma_{12}}{r_{22}} + \frac{r_{12}q_{12}\gamma_{22}}{r_{22}}\right] q_{12} + \cdots$$

$$\text{5 ops: } q_{12} + \frac{7}{2}O(\varepsilon)q_{12} + 6O(\varepsilon)\left[\frac{\gamma_{12}r_{12}q_{11}}{r_{22}} + \frac{\gamma_{22}r_{12}q_{12}}{r_{22}} + \frac{r_{12}q_{11}}{\gamma_{12}}\right] q_{12} + \cdots$$

$$\text{4 ops: } q_{22} + 4O(\varepsilon)q_{22} + 11O(\varepsilon)\left[\frac{r_{12}q_{12}}{\gamma_{22}} + \frac{r_{12}q_{12}\gamma_{12}}{r_{22}} + \frac{r_{12}q_{12}\gamma_{22}}{r_{22}}\right] q_{22} + \cdots$$

$$\text{5 ops: } q_{22} + \frac{7}{2}O(\varepsilon)q_{22} + 6O(\varepsilon)\left[\frac{r_{12}q_{12}}{\gamma_{22}} + \frac{r_{12}q_{11}\gamma_{12}}{r_{22}} + \frac{r_{12}q_{11}\gamma_{22}}{r_{22}}\right] q_{22} + \cdots$$

The orthogonality of the two columns of Q is stated as $q_{11}q_{12} + q_{21}q_{22} = 0$.

With four and five operations the computed values of $q_{11}q_{12} + q_{21}q_{22}$ are (up to terms of order $O(\varepsilon)$):

4 operations

$$q_{11}q_{12} + \frac{15}{2}O(\varepsilon)q_{11}q_{12} + 11O(\varepsilon)q_{11}q_{12}\left[\frac{r_{12}q_{11}}{\gamma_{12}} + \frac{r_{12}q_{11}\gamma_{12}}{r_{22}} + \frac{r_{12}q_{12}\gamma_{22}}{r_{22}}\right] + \cdots$$
$$+ q_{21}q_{12} + \frac{15}{2}O(\varepsilon)q_{21}q_{22} + 11O(\varepsilon)q_{21}q_{22}\left[\frac{r_{12}q_{12}}{\gamma_{12}} + \frac{r_{12}q_{11}\gamma_{12}}{r_{22}} + \frac{r_{12}q_{12}\gamma_{22}}{r_{22}}\right] + \cdots$$

5 operations

$$q_{11}q_{12} + 6O(\varepsilon)q_{11}q_{12} + 6O(\varepsilon)q_{11}q_{12}\left[\frac{r_{12}q_{11}}{\gamma_{12}} + \frac{r_{12}q_{11}\gamma_{12}}{r_{22}} + \frac{r_{12}q_{12}\gamma_{22}}{r_{22}}\right] + \cdots$$
$$+ q_{21}q_{22} + 6O(\varepsilon)q_{21}q_{22} + 6O(\varepsilon)q_{21}q_{22}\left[\frac{r_{12}q_{12}}{\gamma_{12}} + \frac{r_{12}q_{11}\gamma_{12}}{r_{22}} + \frac{r_{12}q_{12}\gamma_{22}}{r_{22}}\right] + \cdots$$

The corresponding spectral information, (sum of square of eigenvalues) $r_{11}^2 + r_{12}^2 + r_{22}^2$ is given by:

4 operations

$$r_{11}^2 + r_{12}^2 + r_{22}^2 + 50\,O(\varepsilon) r_{11}^2 + 130\,O(\varepsilon) r_{12}^2 + 60\,O(\varepsilon) r_{22}^2 + 220\,O(\varepsilon) r_{22} \left[\gamma_{12} r_{12} q_{11} + \gamma_{22} r_{12} q_{12} \right] + \cdots$$

5 operations

$$r_{11}^2 + r_{12}^2 + r_{22}^2 + 30\,O(\varepsilon) r_{11}^2 + 70\,O(\varepsilon) r_{12}^2 + 50\,O(\varepsilon) r_{22}^2 + 120\,O(\varepsilon) r_{22} \left[\gamma_{12} r_{12} q_{11} + \gamma_{22} r_{12} q_{12} \right] + \cdots$$

Even for the case $N = 2$, we can see how rounding degrades spectral information in only one QR decomposition. It is no surprise that the convergence criterion 'gives way' sooner with four operations than with five operations.

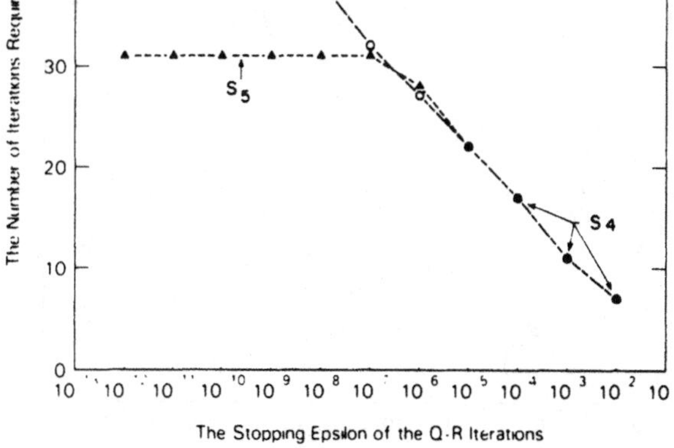

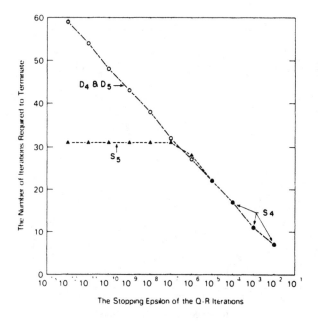

Fig. 4.1: QR Algorithm, $N = 5$

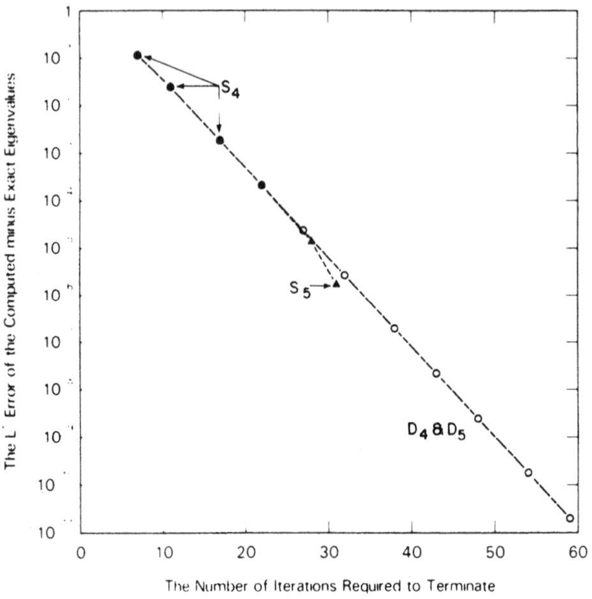

Fig. 4.2: QR Algorithm, $N = 5$

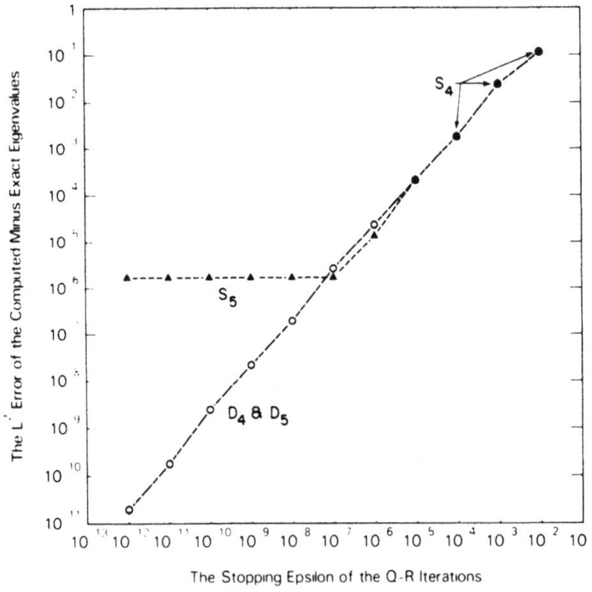

Fig. 4.3: QR Algorithm, $N = 5$

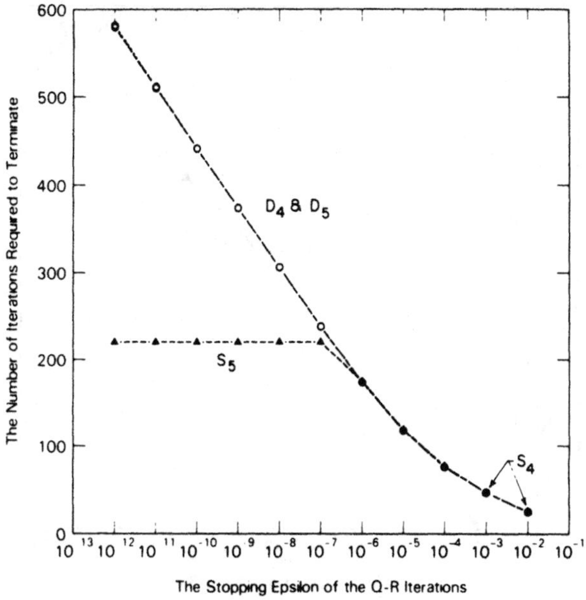

Fig. 4.4: QR Algorithm, $N = 20$

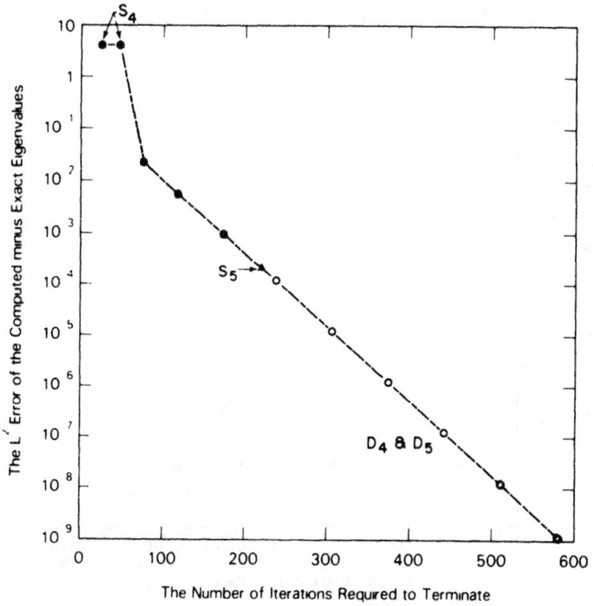

Fig. 4.5: QR Algorithm, $N = 20$

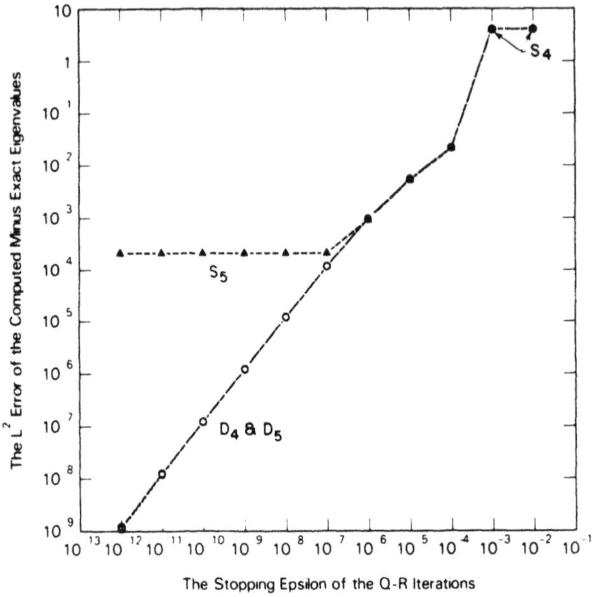

Fig. 4.6: QR Algorithm, $N = 20$

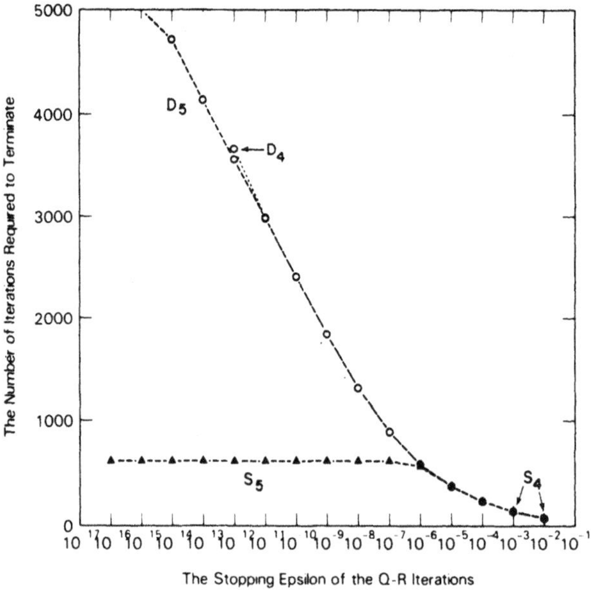

Fig. 4.7: QR Algorithm, $N = 60$

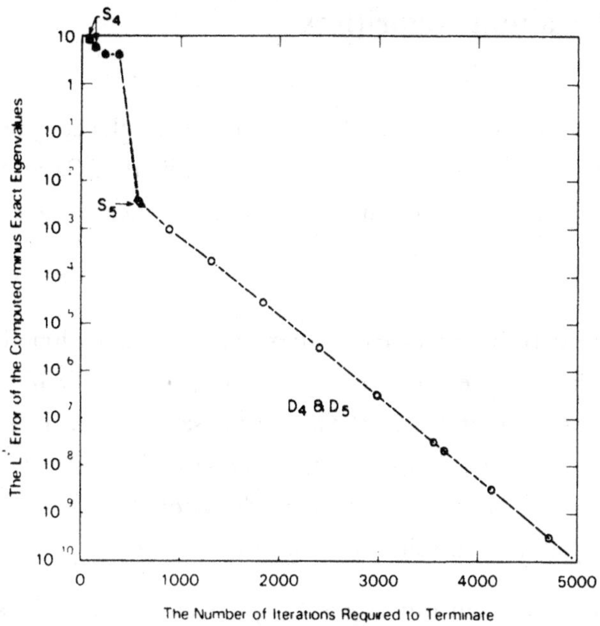

Fig. 4.8: QR Algorithm, $N = 60$

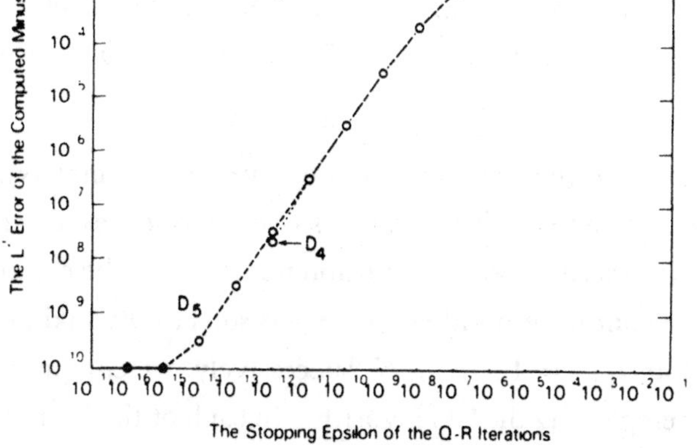

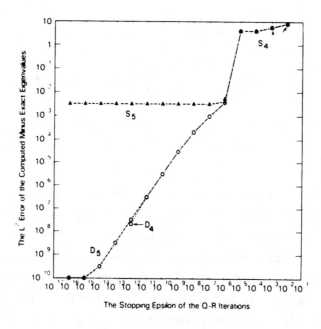

Fig. 4.9: QR Algorithm, $N = 60$

5. Conjugate Gradient Algorithm

The fifth operation has also improved the performance of the conjugate gradient algorithm of Hestenes and Stiefel. The conjugate gradient (CG) algorithm [13] is a well-known iterative method for solving the N-th order linear system

$$Ax = b ,$$

where A is a positive definite symmetric matrix. With exact arithmetic the conjugate gradient algorithm is guaranteed to converge to the exact solution of the linear system in at most n iterations. Qualitatively, the algorithm searches for the solution, moving successively along directions which are conjugate with respect to A. In this way the algorithm eliminates one linearly independent direction from consideration each iteration, necessarily yielding the solution in most N iterations.

The algorithm begins with an initial guess, x_0. It computes and uses the residual $r_0 = b - Ax_0$ as the first search direction, p_0. Each subsequent iteration produces a new x, r and p as follows.

$$a_n = \frac{(r_n, r_n)}{(p_n, Ap_n)}, \quad x_{n+1} = x_n + a_n p_n,$$

$$r_{n+1} = r_n - a_n A p_n, \quad b_n = \frac{(r_{n+1}, r_{n+1})}{(r_n, r_n)}, \quad p_{n+1} = r_{n+1} + b_n p_n.$$

These formulas contain a number of inner products, each of which may be done conventionally with four operations (S_4 or D_4) or more accurately with five operations (S_5 or D_5).

The experiments deal with the five diagonal positive definite matrix A displayed below. The condition number of this matrix A is $O(a^{-2})$ as a goes to zero. By fixing N and allowing a to decrease, a range of condition numbers can be explored. At a given value of a and hence condition number, $Ax = b$ is solved. By taking b to be each of the coordinate vectors e_i, $i = 1, \ldots, N$, successively, we compute the columns of A^{-1}. We do this by employing the CG algorithm in each of the four arithmetic modes S_4, S_5, D_4, D_5. The values computed in each arithmetic mode are:

(i) The number of iterations required by convergence (averaged over the N different problems).

(ii) The Frobenius norm $\| (A^{-1})_{\text{exact}} - (A^{-1})_{\text{computed}} \|_F$. Here $(A^{-1})_{\text{exact}}$ is determined from a known closed form expression for A^{-1}.

$$A = \begin{bmatrix} a^2+1 & -2a & 1 & & & & \\ -2a & a^2+2 & -2a & 1 & & & \\ 1 & -2a & a^2+2 & -2a & 1 & & \\ & 1 & & & \cdots & & \\ & & & & & & 1 \\ & & & & & a^2+2 & -2a \\ & & & & 1 & -2a & a^2+1 \end{bmatrix}$$

Numerical analysts often recommend that a better way to compute the iterates for the CG algorithm is to accumulate the inner products in double precision as the computation proceeds. This mixed precision procedure served as the fifth arithmetic mode (denoted M) that was performed at each condition number value. In the case $N = 9$ and for varying condition number, the values (i) and (ii) resp. are shown in Figs. 5.1 and 5.2 resp. (Figures are found at the end of the section.) Also for a fixed value of parameter a, matrices of various sizes were used in each of the five modes of computation. For $a = 0.01$, these results are shown in Figs. 5.3 and 5.4. An additional set of experiments with A chosen to be a normalized Hilbert segment was also performed. In these experiments, the condition number of A was varied by changing the size of the Hilbert segment, $[\,H\,]_{ij} = 1/(i + j - 1)$. We normalize these segments by multiplying up the elements by an appropriate integer constant to make all the elements integers, and hence computer representable. The results for the Hilbert segments are shown in Figs. 5.5 and 5.6. Even though the CG algorithm is known to be finitely convergent, linear systems with excessively large condition numbers often cause the conjugate gradient algorithm to fail to converge computationally. If a linear system failed to satisfy our convergence criterion (stopping rule), $\| p_n \| < \varepsilon$, after 5000 iterations, that computation was halted.

Three observations can be made. First, S_4 performs poorly when compared to S_5 in both the quality of the solution and the number of iterations required to converge. Next we note that S_5 and M show almost identical results in both quality of the computed inverse and in the mean number of iterations required to converge. Lastly the behavior of D_4 and D_5 are similar. However with increasing problem difficulty D_5 does slightly better than D_4 in both computation quality and mean number of iterations required. In the case of the five diagonal matrix with varying matrix size, the accuracy produced by D_5 is better than D_4 over a wide range by a small yet detectable amount (see Fig. 5.4).

The near equivalent performance of M and S_5 is disarming. It shows that *a computer capable of S_5 computation in the hands of an unsophisticated user can equal the teamwork of a sophisticated numerical analyst and a conventional S_4/D_4 computer.* Indeed the S_5 computer offers other advantages of speed stemming from uniformity (a single S_5 mode rather than a mixed $M = S_4 \cup D_4$ mode).

An explanation of these phenomena comes from considering roundings as a source of noise in the computation. With an extremely difficult problem the amount of noise generated by roundoff at each iteration may well be sufficient to keep the iterates from convergence until a delayed or indefinite time. The rounding counts for the conjugate gradient algorithm are made for the zeroth (start up) iteration and each subsequent iteration. They are:

	4 operations	5 operations
startup	$2N^3$	$2N$
each iteration	$2N^2 + 11N - 1$	$7N + 5$

Upon completion of j iterations past the startup, the rounding counts are:

$$\text{4 ops:} \quad 2N^3 + j\,(2N^2 + 11N - 1)$$
$$\text{5 ops:} \quad 2N + j\,(7N + 5)$$

If j is set to the theoretical maximum value of N, we have $O(N^3)$ roundings for S_4 versus $O(N^2)$ for S_5. See [8, appendix 2] for details.

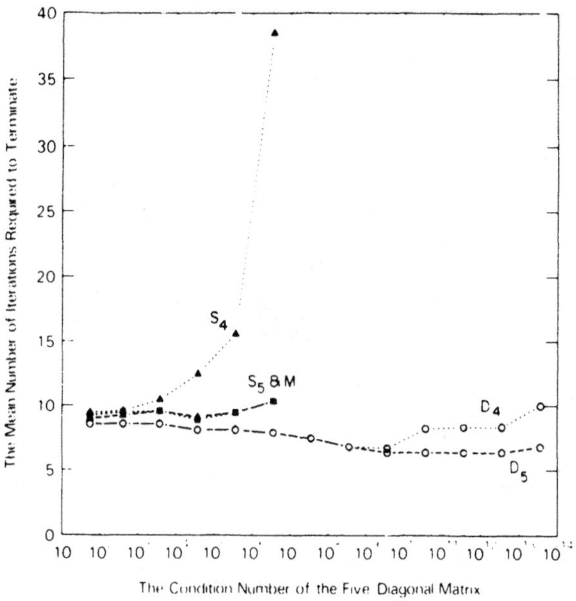

Fig. 5.1: Conjugate Gradient algorithm, 9 × 9 matrix

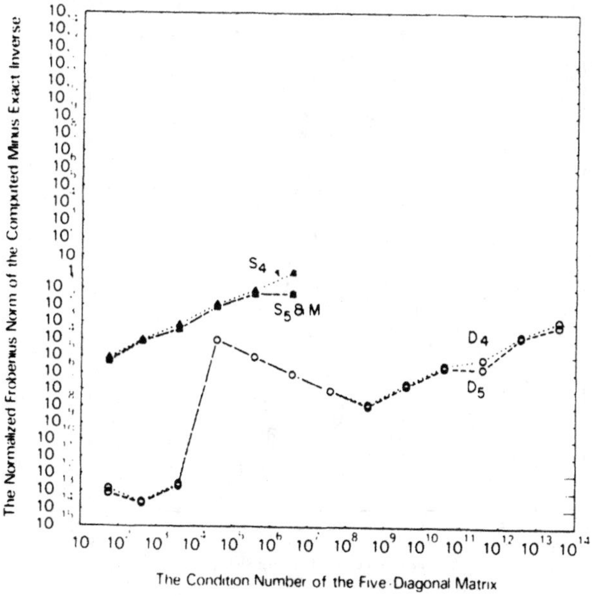

Fig. 5.2: Conjugate Gradient algorithm, 9 × 9 matrix

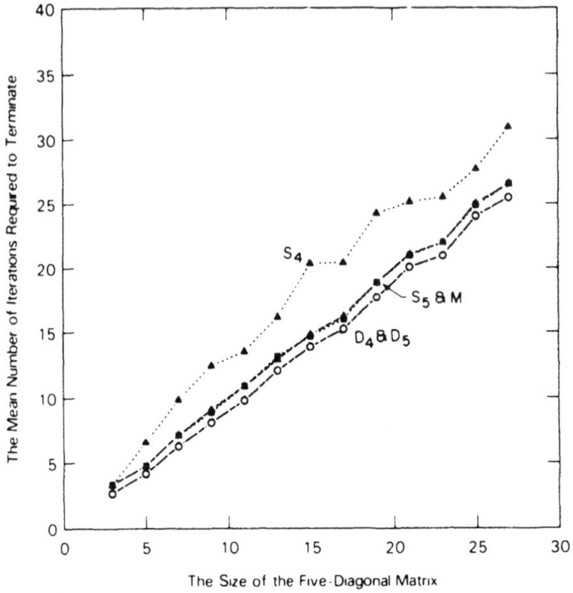

Fig. 5.3: Conjugate Gradient algorithm, $a = 0.01$

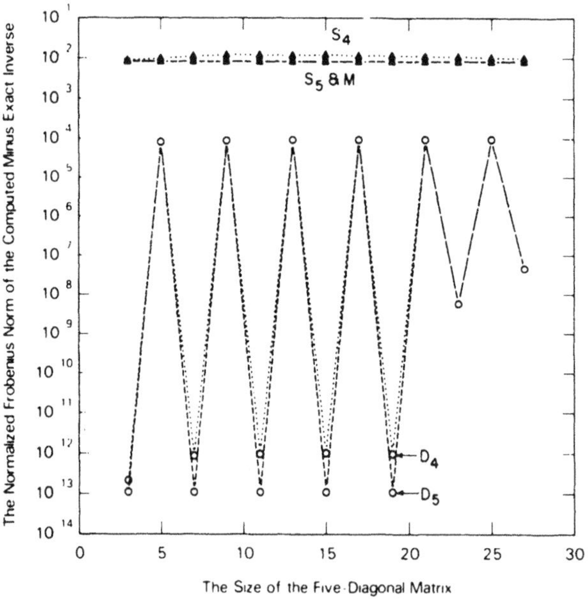

Fig. 5.4: Conjugate Gradient algorithm, $a = 0.01$

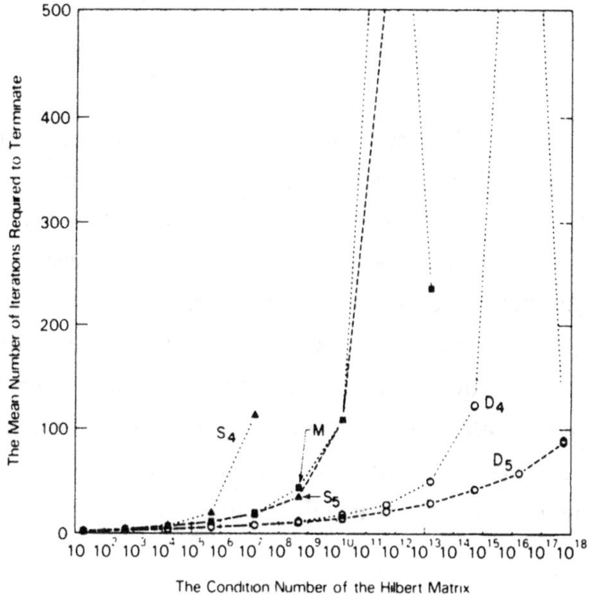

Fig. 5.5: Conjugate Gradient algorithm, Hilbert segments

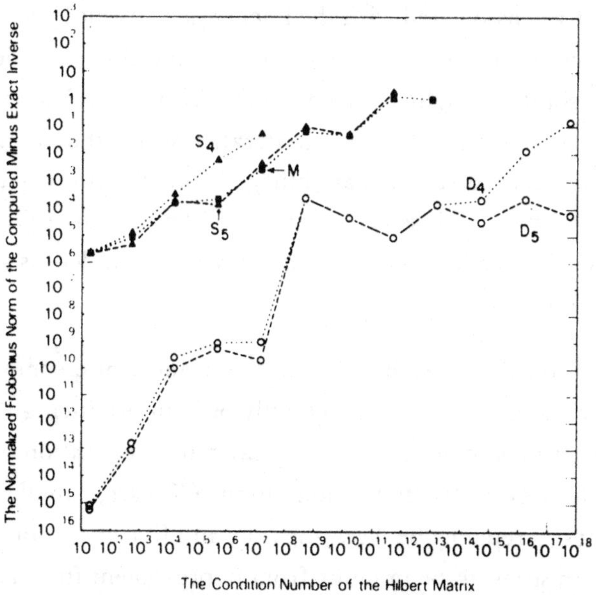

Fig. 5.6: Conjugate Gradient algorithm, Hilbert segments

6. Conclusions

Our study has revealed algorithms susceptible to improvement as well as algorithms which are immune to improvement by high accuracy arithmetic. Those susceptible include (i) the iterative residual correction process applied to refining a numerical approximation to the solution of (poorly conditioned) linear systems such as those which arise in the numerical treatment of ill-posed problems and of stiff differential equations. (ii) the QR algorithm, (iii) the CG algorithm. There are undoubtedly others susceptible to arithmetic improvement. However, our search for susceptible algorithms also uncovered algorithms immune to arithmetic improvement. These immune algorithms included Gaussian elimination (without refinement by residual correction) for solving a linear system and the simplex algorithm for linear programming.

Consider the difference between solving a linear system with the conjugate gradient algorithm and with Gaussian elimination using the Crout algorithm version of the latter. The conjugate gradient algorithm produces a set of conjugate directions along which the quadratic form associated with the linear system is successively minimized. If the conjugate directions are perturbed by noise produced by roundoff, then after N iterations the true solution might not be achieved. However, this state of affairs is detectable, and the CG algorithm continues to iterate towards the solution. In a sense the CG algorithm has a self-correcting capability. Moreover this capability is sensitive to improved arithmetic. This sensitivity is a well-known numerical property for correction procedures such as iterative refinement, which enabled us to generate the studies in sections 2 and 3.

The Crout algorithm, on the other hand, is a direct and finite computation in the sense that the data from A is passed through only once in producing the UL decomposition. Improved arithmetic is wasted by concatenating operations. See [13] for a possible method to circumvent this limitation. In the CG case, cumulative errors will result in a significant discrepancy between the four and five operations performance. The simplex algorithm seems to be immune from improvement for reasons similar to those for the Crout algorithm. A step of the simplex algorithm corresponds to a pivot selection and a step of Gaussian elimination, and indeed there is no possibility offered for a refinment process.

These studies lead us to recommend the inclusion of the accurate inner product as part of the standard floating-point arithmetic available on machines intended for scientific computation. The inner product offers speed-up in the sense that certain iterative algorithms converge faster. Moreover, with hardware implementation of a high speed inner product, many floating-point operations would be replaced with a single operation, thus potentially reducing the time to perform the atomic parts of the computation as well. Note that our studies have not accounted for performance differences attributable to differences in such atomic parts.

The high accuracy inner product also offers the user higher accuracy computation.

Recall that in the QR study, S_5 performed as well as D_4 and D_5 in certain parameter ranges. *Thus another potential time saving aspect is to perform computations in single precision* (S_5), *instead of double precision* (D_4) *thus saving time used in paging information in and out of the main computer memory frequently accompanying double precision computation.*

We speculate that the algorithms discussed here are but a small subset of algorithms susceptible to arithmetic improvement with the accurate inner product. The CG algorithm is finitely convergent in exact arithmetic. However, all convergent iterative algorithms such as QR are finitely convergent with an ε-stopping rule of some sort. The finite convergence property of these iterative algorithms seems to play a role in their susceptability to improvement by means of better quality arithmetic. Thus some other possibilities for susceptible algorithms may be found among iterative, finitely convergent algorithms rich in inner products. For example, the variable metric method for non-linear programming. Of course, algorithms originally found to be immune to arithmetic improvement may prove to be susceptible with a more subtle application of the fifth operation.

References

[1] Bertero, M., DeMol, C., Viano, G.A.: *The Stability of Inverse Problems in Optics,* H.P. Baltes, ed., Springer-Verlag, Berlin 1980.

[2] Kaucher, E., and Miranker, W.L.: *Self-Validating Numerics for Function Space Problems,* Academic Press, N.Y. 1984.

[3] Kaucher, E. and Rump, S.: E-Methods for Fixed-Point Equations f(x) = x, *Computing* 28 (1982)31-42.

[4] Kulish, U.W. and Miranker, W.L.: *Computer Arithmetic in Theory and Practice,* Academic Press, N.Y. 1982.

[5] Kulish, U.W. and Miranker, W.L., eds.: *A New Approach to Scientific Computation,* Academic Press, N.Y. 1983.

[6] Kulisch, U.W. and Miranker, W.L.: The Arithmetic of the Digital Computer, IBM Research Center Report #RC10580, 1984.

[7] Lapidus, L, Aiken, R.C. and Liu, Y.A.: The Occurrence and Numerical Solution of Physical and Chemical Systems Having Widely Varying Time Constants, *Stiff Differential Equations,* R.A. Willoughby, ed., Plenum Press, N.Y. 1979.

[8] Mascagni, M. and Miranker, W.L.: *Arithmetically Improved Algorithmic Performance,* to appear in Computing.

[9] Miranker, W.L.: *Numerical Methods for Stiff Equations and Singular Perturbation Problems,* Reidel Publishing Co., Dordrecht 1981.

[10] Miranker, W.L. and Rump, S.: *Case Studies for ACRITH,* IBM Research Center Report #RC10249, 1983.

[11] Parlett, B.N.: *The Symmetric Eigenvalue Problem,* Prentice-Hall, N.Y. 1980.

[12] Rump, S.: *Solving Algebraic Problems with High Accuracy,* in [5].

[13] Stettler, H.: *to appear.*

[14] Stoer, J. and Bulirsh, R.: *Introduction to Numerical Analysis,* Springer-Verlag, Berlin 1983.

Strict Optimal Error and Residual Estimates for the Solution of Linear Algebraic Systems by Elimination Methods in High-Accuracy Arithmetic

F. Stummel
FB Mathematik, J.W. Goethe-Universität
Frankfurt am Main, Federal Republic of Germany

Abstract. The paper establishes explicit analytical representations of the errors and residuals of the solutions of linear algebraic systems as functions of the data errors and of the rounding errors of a high-accuracy floating-point arithmetic. On this basis, strict, componentwise, and in first order optimal error and residual estimates are obtained. The stability properties of the elimination methods of Doolittle, Crout, and Gauss are compared with each other. The results are applied to three numerical examples arising in difference approximations, boundary and finite element approximations of elliptic boundary value problems. In these examples, only a modest increase of the accuracy of the solutions is achieved by high-accuracy arithmetic.

Introduction

The paper establishes a rounding error analysis of the compact elimination methods (see [2]) of Doolittle and Crout for the solution of linear algebraic systems in high-accuracy floating-point arithmetic. It is assumed that in such an arithmetic scalar products can be computed exactly and that a long dividend can be divided exactly by a floating-point number before the result is rounded to a single-precision floating-point number. The rounding error analysis yields explicit representations of the errors and residuals of the computed solutions as functions of the data errors and of the rounding errors of the floating-point arithmetic. From these representations strict componentwise optimal error and residual bounds are obtained which are formed by means of condition numbers. In addition, error representations and strict optimal error bounds for the computed pivot elements are derived.

In § 5 the error and residual bounds for the elimination method of Doolittle are compared with those for the method of Crout using the associated a priori condition numbers of the algorithms which are first order approximations of the strict condition numbers. It is shown that the difference both between the error and between the residual stability constants $\omega_i = \sigma_i^R/\sigma_i^D$, $\psi_j = \tau_j^R/\tau_j^D$ of the methods of Doolittle and Crout i bounded in absolute value by 1. Further it is readily seen that the method of Doolittle is always more stable than the method of Gauss. Let us remark in passing that the elimination methods ob Banachiewicz and Cholesky in high-accuracy arithmetic can be analyzed in a completely analogous manner.

The error analysis is illustrated by three selected numerical examples in § 6. In particular, the attainable accuracy of the method of Doolittle in high-accuracy arithmetic is compared with that of Gaussian elimination in standard floating-point arithmetic. Example 1 arises from the five-point difference approximation of a Dirichlet boundary value problem for the Poisson equation with a spectral shift near to the middle of the spectrum (see [5]). The linear system has n = 55 equations and band width 11. Example 2 comes from the automotive industry and is obtained by the boundary element method. The associated linear system of n = 86 equations has a full general matrix with no obvious special structure. Example 3 is the finite element approximation (Bogner-Fox-Schmit-element) of a uniformly loaded long thin cantilever with 32 elements. The linear system is relatively ill-conditioned. It has n = 256 equations and the band width 31.

In these examples only a modest increase of the accuracy of the computed solutions of the linear systems is achieved by high-accuracy arithmetic. The errors of the solutions calculated by the method of Doolittle in high-accuracy arithmetic can attain in Example 1 up to around 1/3, in Example 2 up to about 1/30, and in Example 3 up to approximately 1/10 of those calculated by Gaussian elimination.

The present error analysis is very similar to our error analysis of Gaussian elimination in [7]. Our FORTRAN-programs for the error analysis of Gaussian elimination have been described and explained in detail in [8]. The method of Doolittle in high-accuracy arithmetic can be viewed as a Gaussian elimination method in which certain of the intermediate results are calculated exactly. Therefore, the programs in [8] can readily be adapted to the error analysis of the solution of linear algebraic systems with full or band matrices by the eliminati method of Doolittle with partial pivoting.

1. The method of Doolittle

Let $A = (a_{ik})$ be a nonsingular n-by-n matrix having a triangular factorization LU. We are interested in the solution of the inhomogeneous system of linear algebraic equations

$$(1) \qquad Ax = y \quad : \quad \sum_{k=1}^{n} a_{ik}x_k = y_i, \quad i = 1,\dots,n.$$

We shall denote by $A_1 = (a^1_{ik})$ the n-by-(n+1) matrix (A,y). The factors L, U_1 of the trapezoidal decomposition of A_1 are determined, according to the method of Doolittle, by

$$(2) \qquad \begin{aligned} u_{tk} &= a_{tk} - \sum_{j=1}^{t-1} \ell_{tj}u_{jk}, \qquad k = t,\dots,n+1, \\ \ell_{it} &= (a_{it} - \sum_{j=1}^{t-1} \ell_{ij}u_{jt})/u_{tt}, \qquad i = t+1,\dots,n, \end{aligned}$$

for t = 1,...,n or n-1, respectively. The solution x of the linear system (1) is obtained, as usual, from the triangular linear system $Ux = z$ with $z_i = u_{i,n+1}$ by back substitution.

We consider the algorithm under data perturbations and rounding errors of a high-accuracy floating-point arithmetic which permits the computation of u_{tk}, ℓ_{it} in the form

$$(3) \qquad \begin{aligned} \bar{u}_{tk} &= fl(\bar{a}_{tk} - \sum_{j=1}^{t-1} \bar{\ell}_{tj}\bar{u}_{jk}), \qquad k = t,\dots,n+1, \\ \bar{\ell}_{it} &= fl((\bar{a}_{it} - \sum_{j=1}^{t-1} \bar{\ell}_{ij}\bar{u}_{jt})/\bar{u}_{tt}), \qquad i = t+1,\dots,n. \end{aligned}$$

Using the absolute rounding errors d^u_{tk}, d^ℓ_{it} of the rounding operations, (3) can be written

$$(4) \qquad \begin{aligned} \bar{u}_{tk} &= \bar{a}_{tk} - \sum_{j=1}^{t-1} \bar{\ell}_{tj}\bar{u}_{jk} + d^u_{tk}, \qquad k = t,\dots,n+1, \\ \bar{\ell}_{it} &= (\bar{a}_{it} - \sum_{j=1}^{t-1} \bar{\ell}_{ij}\bar{u}_{jt})/\bar{u}_{tt} + d^\ell_{it}, \qquad i = t+1,\dots,n, \end{aligned}$$

for t = 1,...,n or n-1, respectively. Let

$$(5) \qquad \bar{\ell}_{tt} = 1, \quad \bar{\ell}_{it} = \bar{u}_{tk} = 0, \quad i<t,\ k<t.$$

Then the relations (4) imply

(6) $\sum_{j=1}^{n} \bar{\ell}_{ij}\bar{u}_{jt} = \bar{a}_{it} + \bar{u}_{tt}d^{\ell}_{it}, \quad \sum_{j=1}^{n} \bar{\ell}_{tj}\bar{u}_{jk} = \bar{a}_{tk} + d^{u}_{tk},$

for $i>t$, $k\geq t$, and $t = 1,\ldots,n-1$ or n, respectively. Denote by $F_1 = (F_{ik})$ the n-by-(n+1) matrix

(7) $F_{ik} = \bar{u}_{kk}d^{\ell}_{ik}, \; i>k; \quad F_{ik} = d^{u}_{ik}, \; i\leq k.$

The first basic result then reads

(8) _The computed trapezoidal factors_ $\bar{L} = (\bar{\ell}_{ik})$, $\bar{U}_1 = (\bar{u}_{ik})$ _satisfy the relation_

$$\bar{L}\bar{U}_1 = \bar{A}_1 + F_1.$$

From (8) one can readily deduce results concerning the errors $\Delta L = \bar{L} - L$, $\Delta U = \bar{U} - U$ of the computed triangular factors $\bar{L},\bar{U}$ of $\bar{A}$. Using the n-by-n matrix F consisting of the first n columns of F_1, we have

(9) $A = LU, \quad (L + \Delta L)(U + \Delta U) = A + \Delta A + F.$

Consequently,

(10) $L^{-1}\Delta L + \Delta U\bar{U}^{-1} = L^{-1}(\Delta A + F)\bar{U}^{-1}.$

Note that $L^{-1}\Delta L$ is a lower triangular matrix having zero diagonal entries and $\Delta U\bar{U}^{-1}$ is upper triangular. Hence the diagonal of $\Delta U\bar{U}^{-1}$ is equal to the diagonal of the right-hand side in (10) which gives the following representation of the _relative errors of the pivot elements_:

(11) $\dfrac{\Delta u_{ii}}{\bar{u}_{ii}} = \sum_{j,k=1}^{i} \ell^{(-1)}_{ij}(\Delta a_{jk} + F_{jk})\bar{u}^{(-1)}_{ki}, \quad i = 1,\ldots,n.$

The linear system (1) is reduced by forward elimination to the triangular linear system

(12) $U_1\hat{x} = 0 \quad : \quad \sum_{k=i}^{n} u_{ik}x_k - z_i = 0, \quad i = 1,\ldots,n,$

where $z_i = u_{i,n+1}$ and

$$\hat{x} = (x,-1) \quad : \quad \hat{x}_i = x_i, \quad i = 1,\ldots,n, \quad \hat{x}_{n+1} = -1.$$

Let us assume that the triangular linear system can be solved in high-accuracy arithmetic according to

(13) $\bar{x}_i = fl((\bar{z}_i - \sum_{k=i+1}^{n} \bar{u}_{ik}\bar{x}_k)/\bar{u}_{ii}), \quad i = n,\dots,1.$

By introducing the absolute rounding errors d_i^x of the rounding operations in (13) we obtain

(14) $\bar{x}_i = (\bar{z}_i - \sum_{k=i+1}^{n} \bar{u}_{ik}\bar{x}_k)/\bar{u}_{ii} + d_i^x, \quad i = n,\dots,1.$

Consequently, the computed solutions $\bar{x}_i$ are exact solutions of the perturbed triangular linear system

(15) $\sum_{k=i}^{n} \bar{u}_{ik}\bar{x}_k = \bar{z}_i + \bar{u}_{ii}d_i^x, \quad i = 1,\dots,n.$

For simplicity of notation let us define the vectors $\hat{\bar{x}} = (\bar{x},-1)$ and

(16) $f = (f_i), \quad f_i = \bar{u}_{ii}d_i^x, \quad i = 1,\dots,n.$

(17) <u>Then the triangular linear system</u> (15) <u>can be written</u>

$$\bar{U}_1\hat{\bar{x}} = f.$$

2. <u>Error and residual estimates for the method of Doolittle</u>

From the results of § 1 one readily infers representations of the errors $\Delta x = \bar{x} - x$ of the computed solution vector $\bar{x}$ and of the associated residual $A\bar{x} - y = A\Delta x$. By 1.(8),

(1) $LU_1 = A_1, \quad (L + \Delta L)(U_1 + \Delta U_1) = A_1 + \Delta A_1 + F_1,$

and thus

(2) $\Delta L\bar{U}_1 + L\Delta U_1 = \Delta LU_1 + \bar{L}\Delta U_1 = \Delta A_1 + F_1.$

Similarly, from $U_1\hat{x} = 0$ and 1.(17) it follows that

(3) $\bar{U}\Delta x + \Delta U_1\hat{x} = U\Delta x + \Delta U_1\hat{\bar{x}} = f.$

These relations yield the following error and residual approximations

$$(4)\qquad \begin{aligned} \Delta x &= -\bar{U}^{-1}\bar{L}^{-1}(\Delta A_1 + F_1)\hat{x} + \bar{U}^{-1}f,\\ A\bar{x} - y &= A\Delta x = -(\Delta A_1 + F_1)\hat{\bar{x}} + \bar{L}f. \end{aligned}$$

For brevity, denote the matrix $\bar{U}^{-1}\bar{L}^{-1}$ by $G = (g_{ik})$. Then

$$(5)\qquad \begin{aligned} \Delta x_i &= \sum_{j=1}^{n} g_{ij}(\Delta y_j - \sum_{k=1}^{n} \Delta a_{jk}x_k - \sum_{k=1}^{j-1} \bar{u}_{kk}d^{\ell}_{jk} - \sum_{k=j}^{n+1} x_k d^{u}_{jk})\\ &\quad + \sum_{k=i}^{n} \bar{u}^{(-1)}_{ik}\bar{u}_{kk}d^{x}_{k},\\ (A\bar{x} - y)_i &= \Delta y_i - \sum_{k=1}^{n} \Delta a_{ik}\bar{x}_k - \sum_{k=1}^{i-1} \bar{u}_{kk}\bar{x}_k d^{\ell}_{ik} - \sum_{k=i}^{n+1} \bar{x}_k d^{u}_{ik}\\ &\quad + \sum_{k=1}^{i} \bar{\ell}_{ik}\bar{u}_{kk}d^{x}_{k}, \qquad i = 1,\dots,n. \end{aligned}$$

The above representations exhibit the dependence of the errors and residuals upon the data errors Δa_{ik}, Δy_i, and the rounding errors d^{ℓ}_{ik}, d^{u}_{ik}, d^{x}_{i}. Note, that in the residual representation the computed $\bar{x}_k$ and in the error representation the exact solutions x_k of the linear system are used.

We now assume that the data errors are bounded by

$$(6)\qquad |\Delta a_{ik}| \le \alpha_{ik}\eta_D, \qquad |\Delta y_i| \le \beta_i\eta_D, \qquad i,k = 1,\dots,n,$$

using nonnegative weights α_{ik},β_i, and a data accuracy η_D. We further assume that the relative rounding errors are uniformly bounded by a rounding accuracy η_R. Then the associated absolute rounding errors are bounded by

$$(7)\qquad |d^{\ell}_{ik}| \le |\bar{\ell}_{ik}|\eta_R, \qquad |d^{u}_{ik}| \le |\bar{u}_{ik}|\eta_R, \qquad |d^{x}_{i}| \le |\bar{x}_i|\eta_R.$$

If the elements $\bar{a}_{1k}$ of the first row of the perturbed matrix $\bar{A}_1$ are already floating-point numbers, simply $\bar{u}_{1k} = \bar{a}_{1k}$ so that

$$(8)\qquad d^{u}_{1k} = 0, \quad k = 1,\dots,n+1.$$

In the following, the notations

$$(9)\qquad \begin{aligned} &c_{1k} = 0; \quad c_{ik} = |\bar{\ell}_{ik}\bar{u}_{kk}|,\ i>k; \quad c_{ik} = |\bar{u}_{ik}|,\ i\le k;\\ &d_1 = 0; \quad d_i = |\bar{z}_i|,\ i>1; \quad v_i = |\bar{u}_{ii}\bar{x}_i|; \end{aligned}$$

for i,k = 1,...,n are used. Then, under the above assumptions (7), (8), the estimates

(10) $|F_{ik}| \le c_{ik}\eta_R$, $|F_{i,n+1}| \le d_i \eta_R$, $|f_i| \le v_i \eta_R$,

hold for i,k = 1,...,n. The following first error estimate is an immediate consequence of the representation 1.(11):

(11) <u>The relative errors of the computed pivot elements</u> $\bar{u}_{ii}$ <u>are bounded by</u>

$$\left|\frac{\Delta u_{ii}}{\bar{u}_{ii}}\right| \le \sum_{j,k=1}^{i} |\ell_{ij}^{(-1)}|(\alpha_{jk}\eta_D + c_{jk}\eta_R)|\bar{u}_{ki}^{(-1)}|, \quad i = 1,\ldots,n.$$

Next we can readily obtain error and residual estimates for the method of Doolittle. Let us define the vectors of <u>a posteriori residual condition numbers</u> τ_i^D with respect to data perturbations and τ_i^R with respect to rounding errors of the floating-point operations by

(12) $\tau^D = \alpha|\bar{x}| + \beta \quad : \quad \tau_i^D = \sum_{k=1}^{n} \alpha_{ik}|\bar{x}_k| + \beta_i, \quad i = 1,\ldots,n,$

and, in vector notation,

(13) $\tau^R = \tau^o + \tau^1, \quad \tau^o = C|\bar{x}| + d, \quad \tau^1 = |\bar{L}|v.$

Additionally, let the vector of <u>total residual condition numbers</u> be defined by

(14) $\tau = \frac{\eta_D}{\eta}\tau^D + \frac{\eta_R}{\eta}\tau^R, \quad \eta = \max(\eta_D,\eta_R).$

From (5) and under the assumptions (6), (7) one infers the following result for the method of Doolittle:

(15) <u>The residual of the computed solution vector</u> $\bar{x}$ <u>of the linear system satisfies the optimal componentwise estimate</u>

$$|A\bar{x} - y| \le \eta\tau.$$

For the corresponding error estimates we introduce the following vectors σ^D, σ^R of <u>absolute data and rounding condition numbers</u>

$$\text{(16)}\quad \begin{aligned} \sigma^D &= |\bar{U}^{-1}\bar{L}^{-1}|\tau^D, \\ \sigma^R &= \sigma^o + \sigma^1, \quad \sigma^o = |\bar{U}^{-1}\bar{L}^{-1}|\tau^o, \quad \sigma^1 = |\bar{U}^{-1}|v. \end{aligned}$$

In addition, the vector of <u>absolute total condition numbers</u>

$$\text{(17)}\quad \sigma = \frac{\eta_D}{\eta}\sigma^D + \frac{\eta_R}{\eta}\sigma^R$$

is needed. Finally, let the <u>relative errors</u> of $\bar{x}_i$ and the associated <u>relative total condition numbers</u> be defined by

$$\text{(18)}\quad Px_i = \frac{\bar{x}_i - x_i}{\bar{x}_i}, \qquad \rho_i = \frac{\sigma_i}{|\bar{x}_i|} \quad (\bar{x}_i \neq 0), \qquad i = 1,\ldots,n,$$

and the <u>maximal relative condition number</u> by

$$\text{(19)}\quad \rho_m = \max_{i=1,\ldots,n} \rho_i .$$

The representation (5) and the above assumptions (6), (7) then imply the following error estimates (see [7]).

(20) <u>If all</u> $\bar{x}_i$ <u>are different from zero and</u> η <u>is so small that</u> $\rho_m\eta<1$, <u>the absolute and relative errors of the computed solution vector</u> $\bar{x}$ <u>are bounded componentwise and in first order optimally by</u>

$$|\Delta x| \leq \frac{1}{1-\rho_m\eta}\,\eta\sigma, \qquad |Px| \leq \frac{1}{1-\rho_m\eta}\,\eta\rho .$$

The residual condition numbers are simple functions of the data and of the elements of the trapezoidal factors of $\bar{A}$. The computation of the condition numbers σ_i^D, σ_i^o, σ_i^1 requires the computation of the i-th rows of $\bar{U}^{-1}$ and $\bar{U}^{-1}\bar{L}^{-1}$ which will be analyzed in the next section. Let us remark in passing that one can very easily determine <u>lower bounds</u> for the vectors of condition numbers σ^D, σ^o, σ^1 by calculating the solutions of the linear systems

$$\text{(21)}\quad \bar{L}\bar{U}s^D = \tau^D, \qquad \bar{L}\bar{U}s^o = \tau^o, \qquad \bar{U}s^1 = v,$$

where $\bar{L},\bar{U}$ are the computed triangular factors of $\bar{A}$. Then, evidently, the componentwise estimates

(22) $|s^D| \le |\bar{U}^{-1}\bar{L}^{-1}|\tau^D = \sigma^D, \quad |s^o| \le |\bar{U}^{-1}\bar{L}^{-1}|\tau^o = \sigma^o, \quad |s^1| \le |\bar{U}^{-1}|v = \sigma^1,$

hold. The more $\bar{U}^{-1}\bar{L}^{-1}$ and $\bar{U}^{-1}$ are nonnegative the better σ^D, σ^o, σ^1 are approximated by the lower bounds $|s^D|$, $|s^o|$, $|s^1|$, respectively.

3. A posteriori condition numbers

The residual condition numbers τ_i^D, τ_i^R in § 2 are simple arithmetic expressions consisting solely of sums and products of nonnegative weights, of absolute values of computed intermediate results of the elimination method, and of the computed solutions $\bar{x}_i$. Hence the numerical evaluation of the residual condition numbers is very stable. In contrast, the evaluation of the condition numbers σ_i^D, σ_i^R requires the computation of the i-th rows of the matrices $H = \bar{U}^{-1}$, $G = \bar{U}^{-1}\bar{L}^{-1}$. Let $\bar{H} = (\bar{h}_{ik})$, $\bar{G} = (\bar{g}_{ik})$ denote the approximations of $H = (h_{ik})$, $G = (g_{ik})$, computed in high-accuracy arithmetic. Define the vectors of absolute a posteriori data and rounding condition numbers by

$$\text{(1)} \qquad \begin{aligned} \bar{\sigma}^D &= |\bar{G}|\tau^D, \\ \bar{\sigma}^R &= \bar{\sigma}^o + \bar{\sigma}^1, \quad \bar{\sigma}^o = |\bar{G}|\tau^o, \quad \bar{\sigma}^1 = |\bar{H}|v. \end{aligned}$$

The associated vector of absolute total a posteriori condition numbers reads

$$\text{(2)} \qquad \bar{\sigma} = \frac{\eta_D}{\eta}\bar{\sigma}^D + \frac{\eta_R}{\eta}\bar{\sigma}^R, \qquad \eta = \max(\eta_D, \eta_R).$$

The a posteriori condition numbers are arithmetic expressions built-up completely of sums and products of absolute values of computed quantities.

The vectors $\bar{\sigma}^D$, $\bar{\sigma}^R$, $\bar{\sigma}$ are first order approximations of the vectors σ^D, σ^R, σ, respectively, of the strict condition numbers defined in § 2. We can readily obtain a bound for the error of the approximation $\bar{\sigma}$ of σ

as we will show now. Since $H = \bar{U}^{-1}$ is upper triangular and satisfies the system of equations

$$(3) \quad H\bar{U} = I \quad : \quad \sum_{j=i}^{k} h_{ij}\bar{u}_{jk} = \delta_{ik}, \quad i,k = 1,\dots,n,\ k \geq i,$$

the entries of the i-th row of $\bar{H}$ are calculated according to

$$\begin{aligned} &\bar{h}_{i1} = \dots = \bar{h}_{i,i-1} = 0, \\ (4) \quad &\bar{h}_{ii} = fl\left(\frac{1}{\bar{u}_{ii}}\right) = \frac{1}{\bar{u}_{ii}} + d^h_{ii}, \\ &\bar{h}_{ik} = fl\left(-\left(\sum_{j=i}^{k-1} \bar{h}_{ij}\bar{u}_{jk}\right)/\bar{u}_{kk}\right) = -\left(\sum_{j=i}^{k-1} \bar{h}_{ij}\bar{u}_{jk}\right)/\bar{u}_{kk} + d^h_{ik} \end{aligned}$$

where the d^h_{ik} denote the absolute rounding errors occurring in (4). The equations (4) can be rewritten in the form

$$(5) \quad \bar{H}\bar{U} = I + F_1$$

where

$$F_1 = (f^1_{ik}), \quad f^1_{ik} = \bar{u}_{kk} d^h_{ik}, \quad i,k = 1,\dots,n,$$

with $d^h_{ik} = 0$ for $k<i$.

The matrix $G = \bar{U}^{-1}\bar{L}^{-1}$ satisfies the relation

$$(6) \quad G\bar{L} = H \quad : \quad \sum_{j=k}^{n} g_{ij}\bar{\ell}_{jk} = h_{ik}, \quad i,k = 1,\dots,n.$$

Thus the entries of the i-th row of $\bar{G}$ can be computed by

$$\begin{aligned} &\bar{g}_{in} = \bar{h}_{in}, \\ (7) \quad &\bar{g}_{ik} = fl\left(\bar{h}_{ik} - \sum_{j=k+1}^{n} \bar{g}_{ij}\bar{\ell}_{jk}\right) = \bar{h}_{ik} - \sum_{j=k+1}^{n} \bar{g}_{ij}\bar{\ell}_{jk} + d^g_{ik} \end{aligned}$$

for $k = n-1,\dots,1$. Consequently,

$$(8) \quad \bar{G}\bar{L} = \bar{H} + F_2$$

with

$$F_2 = (f^2_{ik}), \quad f^2_{ik} = d^g_{ik}, \quad i,k = 1,\dots,n,$$

and $d^g_{in} = 0$.

Let us assume as in § 2 that the absolute rounding errors d_{ik}^{h}, d_{ik}^{g} are bounded by

(9) $|d_{ik}^{h}| \leq |\bar{h}_{ik}|\eta_R; \quad |d_{ik}^{g}| \leq |\bar{g}_{ik}|\eta_R, \quad k<n;$

and put

$$(10) \quad \begin{aligned} \Phi_1 &= (\varphi_{ik}^1) \quad : \quad \varphi_{ik}^1 = |\bar{h}_{ik}\bar{u}_{kk}|, \quad i,k = 1,\ldots,n, \\ \Phi_2 &= (\varphi_{ik}^2) \quad : \quad \varphi_{ik}^2 = |\bar{g}_{ik}|,\ k<n; \quad \varphi_{in}^2 = 0. \end{aligned}$$

Then the residual matrices F_1, F_2 in (5), (8) are bounded componentwise by

(11) $|F_1| \leq \Phi_1\eta_R, \quad |F_2| \leq \Phi_2\eta_R.$

In the following error estimate we use the weighted maximum norm

(12) $||x|| = ||\Lambda^{-1}x||_\infty = \max\limits_{i=1,\ldots,n} |\frac{x_i}{\lambda_i}|, \quad x \in \mathbb{R}^n,$

where $\Lambda = \mathrm{diag}(\lambda_1,\ldots,\lambda_n)$ using weights $\lambda_i \neq 0$. An associated norm for the n-by-n matrices $A = (a_{ik})$, compatible with this vector norm, is given by

(13) $||A|| = ||\Lambda^{-1}A\Lambda||_\infty = \max\limits_{i=1,\ldots,n} \frac{1}{|\lambda_i|} \sum\limits_{k=1}^{n} |a_{ik}\lambda_k|$

(14) _Using the matrices_ Φ_1, Φ_2 _in_ (10) _and_ $\Phi_o = \Phi_1 + \Phi_2|\bar{U}|$, _let_ η_R _be so small that_

$$\max(||\Phi_o||, ||\Phi_1||) < \frac{1}{\eta_R} .$$

Then the error estimate

$$||\bar{\sigma} - \sigma|| \leq \gamma\eta_R$$

is valid with the constants

$$\gamma = \varphi_o || \frac{\eta_D}{\eta} \bar{\sigma}^D + \frac{\eta_R}{\eta} \bar{\sigma}^o || + \varphi_1 \frac{\eta_R}{\eta} || \bar{\sigma}^1 ||$$

and

$$\varphi_o = \frac{||\Phi_o||}{1-||\Phi_o||\eta_R}, \quad \varphi_1 = \frac{||\Phi_1||}{1-||\Phi_1||\eta_R} .$$

The proof of this theorem is found in [7]. By choosing the weights $\lambda_i = 1$, the above theorem gives an error estimate for the absolute condition numbers:

(15) $|\bar{\sigma}_i - \sigma_i| \le \gamma_1 \eta_R, \quad i = 1,\dots,n.$

When all computed solutions $\bar{x}_i$ are different from zero, one can put $\lambda_i = |\bar{x}_i|$ and obtains estimates of the form

(16) $|\bar{\rho}_i - \rho_i| \le \gamma_2 \eta_R, \quad i = 1,\dots,n,$

for the relative condition numbers. In this case,

$$||\bar{\sigma}^D|| = ||\bar{\rho}^D||_\infty, \quad ||\bar{\sigma}^o|| = ||\bar{\rho}^o||_\infty, \quad ||\bar{\sigma}^1|| = ||\bar{\rho}^1||_\infty,$$

using the <u>relative a posteriori condition numbers</u>

(17) $\bar{\rho}_i^D = \dfrac{\bar{\sigma}_i^D}{|\bar{x}_i|}, \quad \bar{\rho}_i^o = \dfrac{\bar{\sigma}_i^o}{|\bar{x}_i|}, \quad \bar{\rho}_i^1 = \dfrac{\bar{\sigma}_i^1}{|\bar{x}_i|}, \quad i = 1,\dots,n.$

Another suitable choice of the weights could be $\lambda_i = \bar{\sigma}_i^D$ which gives

(18) $|\bar{\sigma}_i - \sigma_i| \le \bar{\sigma}_i^D \gamma_3 \eta_R, \quad i = 1,\dots,n.$

In the last case,

$$||\bar{\sigma}^D|| = 1, \quad ||\bar{\sigma}^o|| = ||\omega^o||_\infty, \quad ||\bar{\sigma}^1|| = ||\omega^1||_\infty,$$

using the <u>stability constants</u>

(19) $\omega_i^o = \dfrac{\bar{\sigma}_i^o}{\bar{\sigma}_i^D}, \quad \omega_i^1 = \dfrac{\bar{\sigma}_i^1}{\bar{\sigma}_o^D}, \quad i = 1,\dots,n.$

4. The method of Crout

The method of Crout (see [1],[2]) determines the trapezoidal factors L, U_1 of the matrix A_1 by means of the formulas

(1)
$$\ell_{it} = a_{it} - \sum_{j=1}^{t-1} \ell_{ij} u_{jt}, \quad i = t,\dots,n,$$
$$u_{tk} = (a_{tk} - \sum_{j=1}^{t-1} \ell_{tj} u_{jk}) / \ell_{tt}, \quad k = t+1,\dots,n+1,$$

for $t = 1,\dots,n$. Now the diagonal entries u_{tt} of U_1 are equal to 1.

Under data perturbations and in a high-accuracy floating-point arithmetic

$$\text{(2)}\qquad \begin{aligned} \bar{\ell}_{it} &= fl(\bar{a}_{it} - \sum_{j=1}^{t-1} \bar{\ell}_{ij}\bar{u}_{jt}), \quad i = t,\dots,n, \\ \bar{u}_{tk} &= fl((\bar{a}_{tk} - \sum_{j=1}^{t-1} \bar{\ell}_{tj}\bar{u}_{jk})/\bar{\ell}_{tt}), \quad k = t+1,\dots,n+1. \end{aligned}$$

In addition, let

$$\text{(3)}\qquad \bar{u}_{tt} = 1, \quad \bar{\ell}_{it} = \bar{u}_{tk} = 0, \quad i<t,\ k<t.$$

Then the representations

$$\text{(4)}\qquad \sum_{j=1}^{n} \bar{\ell}_{ij}\bar{u}_{jt} = \bar{a}_{it} + d^{\ell}_{it}, \quad \sum_{j=1}^{n} \bar{\ell}_{tj}\bar{u}_{jk} = \bar{a}_{tk} + \bar{\ell}_{tt}d^{u}_{tk},$$

hold for $i\geq t$, $k>t$, where d^{ℓ}_{it}, d^{u}_{tk} are the absolute errors of the rounding functions in (2). If $\bar{a}_{i1} = fl(\bar{a}_{i1})$, there are no rounding errors in determining $\bar{\ell}_{i1}$ and thus

$$\text{(5)}\qquad d^{\ell}_{i1} = 0, \quad i = 1,\dots,n.$$

(6) _Let the_ n-_by_-(n+1) _perturbation matrix_ $F_1 = (F_{ik})$ _be defined by_

$$F_{ik} = d^{\ell}_{ik}, \quad i\geq k; \qquad F_{ik} = \bar{\ell}_{ii}d^{u}_{ik}, \quad i<k.$$

Then

$$\bar{L}\bar{U}_1 = \bar{A}_1 + F_1.$$

The triangular linear system in the method of Crout reads

$$\text{(7)}\qquad x_i + \sum_{k=i+1}^{n} u_{ik}x_k = z_i, \quad i = 1,\dots,n,$$

which is solved in high-accuracy arithmetic by

$$\text{(8)}\qquad \bar{x}_i = fl(\bar{z}_i - \sum_{k=i+1}^{n} \bar{u}_{ik}\bar{x}_k), \quad i = n,\dots,1.$$

Let d^{x}_{i} denote the associated absolute rounding errors.

(9) _Then_

$$\bar{U}_1\hat{x} = f$$

using the right-hand side

$$f = (f_i), \quad f_i = d^{x}_{i}, \quad i = 1,\dots,n.$$

From the above results and 2.(4) one readily infers the <u>error and residual representations</u>

(10)
$$\Delta x_i = \sum_{j=1}^{n} g_{ij}\Big(\Delta y_j - \sum_{k=1}^{n} \Delta a_{jk}x_k - \sum_{j=1}^{i} d^{\ell}_{jk}x_k - \sum_{k=j+1}^{n+1} \bar{\ell}_{jj}d^{u}_{jk}x_k\Big) + \sum_{k=i}^{n} \bar{u}^{(-1)}_{ik} d^{x}_{k};$$

$$(A\bar{x} - y)_i = \Delta y_i - \sum_{k=1}^{n} \Delta a_{ik}\bar{x}_k - \sum_{k=1}^{i} d^{\ell}_{ik}\bar{x}_k - \sum_{k=i+1}^{n+1} \bar{\ell}_{ii}d^{u}_{ik}\bar{x}_k + \sum_{k=1}^{i} \bar{\ell}_{ik}d^{x}_{k}, \quad i = 1,\dots,n.$$

Now put

(11)
$$c_{i1} = 0; \quad c_{ik} = |\bar{\ell}_{ik}|, \quad i \geq k > 1; \quad c_{ik} = |\bar{\ell}_{ii}\bar{u}_{ik}|, \quad i<k;$$
$$d_i = |\bar{\ell}_{ii}\bar{z}_i|; \quad v_i = |\bar{x}_i|;$$

for $\bar{z}_i = \bar{u}_{i,n+1}$, $i,k = 1,\dots,n$. Then the following theorem holds.

(12) <u>Let the error and residual condition numbers</u> σ^D_i, σ^R_i, σ_i, τ^D_i, τ^R_i, τ_i <u>be defined as in § 2 using the above coefficients</u> (11). <u>Then the error and residual estimates</u> 2.(15), (20) <u>hold for the solution vector</u> $\bar{x}$ <u>of the method of Crout.</u>

5. Comparison of the elimination methods of Gauss, Doolittle, and Crout

In order to compare the error and residual bounds of different methods for the solution of a given problem we use the associated a priori error and residual condition numbers. These are obtained from the strict condition numbers defined in § 2 by replacing all perturbed or computed quantities $\bar{a}_{ik}, \bar{\ell}_{ik}, \bar{u}_{ik}, \bar{x}_k$ by the corresponding exact quantities $a_{ik}, \ell_{ik}, u_{ik}, x_k$. The a priori and the strict condition numbers differ only by terms of first order in the accuracy constant $\eta = \max(\eta_D, \eta_R)$.

For a comparison of the elimination methods of Gauss, Doolittle, and Crout in this section we shall denote by L,U the triangular factors of A and by z the right-hand side of the triangular linear system defined in § 1. Thus

$$(1)\qquad \begin{aligned} &\ell_{i1}u_{11} = a_{i1}, \quad \ell_{ii} = 1, \quad i = 1,\dots,n,\\ &u_{1k} = a_{1k}, \quad k = 1,\dots,n. \end{aligned}$$

For the algorithm of Crout the triangular factors L^C, U^C of A and the right-hand side z^C then have the representation

$$(2)\qquad \ell_{ik}^C = \ell_{ik}u_{kk}, \quad u_{ik}^C = \frac{u_{ik}}{u_{ii}}, \quad z_i^C = \frac{z_i}{u_{ii}}, \quad i,k = 1,\dots,n.$$

The <u>a priori residual condition numbers</u> τ_i^o, τ_i^1 of the <u>method of Doolittle</u> read

$$(3i)\quad \tau_1^o = 0, \quad \tau_i^o = \sum_{k=1}^{i-1} |\ell_{ik}u_{kk}x_k| + \sum_{k=i}^{n} |u_{ik}x_k| + |z_i|,$$

for i = 2,...,n and

$$(3ii)\quad \tau_i^1 = \sum_{k=1}^{i} |\ell_{ik}u_{kk}x_k|, \quad i = 1,\dots,n.$$

The <u>a priori residual condition numbers</u> of the <u>method of Crout</u> have in these notations the form

$$(4)\qquad \begin{aligned} &\tau_i^o = \sum_{k=2}^{i} |\ell_{ik}u_{kk}x_k| + \sum_{k=i+1}^{n} |u_{ik}x_k| + |z_i|,\\ &\tau_i^1 = \sum_{k=1}^{i} |\ell_{ik}u_{kk}x_k|, \quad i = 1,\dots,n. \end{aligned}$$

Consequently,

$$(5)\qquad |\text{Crout-}\tau_i^o - \text{Doolittle-}\tau_i^o| = \delta_i, \quad i = 1,\dots,n,$$

with

$$\delta_1 = \sum_{k=2}^{n} |a_{1k}x_k| + |y_1|, \quad \delta_i = |a_{i1}x_1|, \quad i = 2,\dots,n.$$

The vectors σ^D, τ^D of the a priori error and residual condition numbers with respect to relatively uniformly bounded data perturbations are specified by

(6) $\tau^D = |A||x| + |y|, \quad \sigma^D = |A^{-1}|\tau^D.$

Hence $\delta_i \le \tau_i^D$ and so

(7) $|\text{Crout-}\tau_i^o - \text{Doolittle-}\tau_i^o| \le \tau_i^D, \quad |\text{Crout-}\sigma_i^o - \text{Doolittle-}\sigma_i^o| \le \sigma_i^D.$

Since the a priori condition numbers τ_i^1 of the methods of Doolittle and Crout are the same it follows from the above that

(8) $|\text{Crout-}\tau_i^R - \text{Doolittle-}\tau_i^R| \le \tau_i^D, \quad |\text{Crout-}\sigma_i^R - \text{Doolittle-}\sigma_i^R| \le \sigma_i^D.$

The method of Doolittle can be viewed as a Gaussian elimination method in a high-accuracy arithmetic. For, the formulas 1.(3) are equivalent to

(9i)
$$\bar{u}_{tk} = fl(\bar{a}_{tk}^t),\ k = t,\dots,n+1; \quad \bar{\ell}_{it} = fl\left(\frac{\bar{a}_{it}^t}{\bar{u}_{tt}}\right), \quad i = t+1,\dots,n;$$
$$\bar{a}_{ik}^{t+1} = \bar{a}_{ik}^t - \bar{\ell}_{it}\bar{u}_{tk}, \quad i = t+1,\dots,n,\ k = t+1,\dots,n+1;$$

for $t = 1,\dots,n-1$, and

(9ii) $u_{nk} = fl(\bar{a}_{nk}^n), \quad k = n,n+1.$

Analogously, back substitution 1.(13) can be written in the form

(10i) $\bar{z}_i^{n+1} = \bar{z}_i, \quad \bar{z}_i^k = \bar{z}_i^{k+1} - \bar{u}_{ik}\bar{x}_k, \quad k = n,\dots,i+1,$

and

(10ii) $\bar{x}_i = fl\left(\frac{\bar{z}_i^{i+1}}{\bar{u}_{ii}}\right), \quad i = n,\dots,1.$

In Gaussian elimination in standard floating-point arithmetic also the multiplications and subtractions in (9), (10) are carried out with a rounding operation. The above considerations readily yield the result that our a priori error and residual condition numbers for Gaussian elimination in [4], [5], [8] are greater than or equal to those of the Doolittle method. That is,

(11) $\text{Doolittle-}\tau_i \le \text{Gauss-}\tau_i, \quad \text{Doolittle-}\sigma_i \le \text{Gauss-}\sigma_i,$

for $\tau_i = \tau_i^o, \tau_i^1, \tau_i^R$ and $\sigma_i = \sigma_i^o, \sigma_i^1, \sigma_i^R$, $i = 1,\dots,n.$

6. Numerical examples

We have tested the first order error and residual estimates for the Doolittle method in a series of numerical examples. For comparison, the corresponding error and residual estimates for Gaussian elimination in standard floating-point arithmetic were calculated. The numerical examples were computed on a Micro PDP 11 computer. The high-accuracy arithmetic was simulated by a kind of fl_2-arithmetic in the sense of [9]. That is, arithmetic operations were carried out in double precision (56 binary digits) and the results rounded to a prescribed number of digits. As rounding function symmetric rounding, chopping, monotone rounding upwards or downwards, respectively, could be chosen.

In all examples the coefficients and the components of the right-hand sides were rounded to the precision of the floating-point arithmetic used in the computation. Then the solutions of the system both in 56 binary digit double precision and in a given N binary digit accuracy were computed. In this way, there are no data perturbations that have to be taken into account. The floating-point accuracy constant η_R has the value 2^{-N} for symmetric rounding and 2^{-N+1} for chopping and monotone rounding. The error and residual percentages

$$P_i\,{}^o/o = 100\,\frac{\bar{x}_i - x_i}{\bar{\sigma}_i^R \eta_R}\,, \qquad Q_i\,{}^o/o = 100\,\frac{(A\bar{x}-y)_i}{\tau_i^R \eta_R}\,, \qquad i = 1,\dots,n,$$

measure to which extent our error and residual estimates are realistic.

<u>Example 1</u> (see [5]) is obtained by the well-known five-point difference approximation of the Poisson equation with zero boundary conditions using an equidistant mesh of mesh width 1 in a rectangle having side lengths 12 and 6. The matrix A of the system consists of $n = 55$ rows and columns and has band width 11. In addition, a spectral shift $\lambda = 3.99$ near to the middle of the spectrum of A was used. The

Table 1. Five-point difference approximation of the Poisson equation, condition numbers and stability constants

		without pivoting		with pivoting		
		Gauss	Doolittle	Gauss	Doolittle	
i	ρ_i^D	ω_i	ω_i	ω_i	ω_i	x_i
1	7.43+02	11.8	2.45	3.34	1.17	-.515
2	5.77+02	73.0	19.9	3.17	1.13	-.500
3	6.74+02	12.3	3.12	3.61	1.16	-.515
4	5.96+02	61.1	15.6	2.89	1.09	-.500
5	7.42+02	9.19	2.06	2.80	1.03	-.515
6	5.89+02	72.8	18.8	3.16	1.10	-.505
7	1.49+04	8.32	1.93	3.28	1.19	.026
8	4.56+04	63.6	12.6	2.83	1.32	-.005
9	1.47+04	9.46	2.37	3.01	1.04	.026
10	5.68+04	.611	.163	.030	.011	-.505

		Gauss	Doolittle	Gauss	Doolittle
i	τ_i^D	ψ_i	ψ_i	ψ_i	ψ_i
1	2.00	.755	2.57-03	51.0	5.40
2	2.05	394	123	1.51	.250
3	2.00	3.30	1.02	1.50	1.25
4	2.06	122	37.9	1.50	1.25
5	2.00	3.49	1.52	1.00	1.25
6	2.05	810	116	56.6	5.80
7	2.02	6.09	.936	8.27	2.26
8	2.10	336	42.3	6.23	2.00
9	2.02	5.10	.718	2.02	.515
10	2.06	206	17.4	1.23	1.25

Table 2. Boundary element method, data condition numbers, stability constants, solutions x_i, associated error and residual percentages

		Gauss		Doolittle		
i	ρ_i^D	ω_i	$P_i\%$	ω_i	$P_i\%$	x_i
40	6.47+01	30.6	16	1.23	13	-5.35-06
41	7.93+01	29.4	13	1.13	8	1.03-05
42	2.18+03	27.7	-1	1.05	-7	-1.40-07
43	9.30+01	30.1	15	1.27	14	-6.25-06
44	5.80+01	31.0	17	1.23	14	-7.72-06
45	1.14+02	29.7	15	1.26	15	5.07-06
46	1.43+02	27.7	17	1.23	15	-2.91-06
47	1.02+02	30.2	16	1.26	15	8.09-06
48	1.30+02	30.2	17	1.25	15	-3.69-06
49	1.89+02	23.2	9	1.29	4	-3.42-06
50	1.00+04	20.2	6	1.29	2	-4.68-08

		Gauss		Doolittle	
i	τ_i^D	ψ_i	$Q_i\%$	ψ_i	$Q_i\%$
40	5.42	35.8	-16	1.18	0
41	6.21	36.9	-11	1.20	16
42	4.02	44.8	- 6	1.37	-18
43	5.94	43.8	-15	1.78	- 7
44	9.20	40.5	-12	1.37	15
45	4.99	42.9	- 3	1.76	9
46	4.95	30.9	-10	1.16	- 2
47	3.85	74.8	- 4	2.91	9
48	4.26	41.8	- 7	1.62	-16
49	4.44	66.7	- 5	2.75	- 6
50	6.79	69.5	- 8	2.21	-14

matrix $A-\lambda E$ and the right-hand side $y = (1,\dots,1)$ were multiplied by .9973 and then rounded symmetrically to 15 binary digits. Table 1 shows some of the residual condition numbers τ_i^D and of the relative condition numbers $\rho_i^D = \sigma_i^D/|\bar{x}_i|$ with respect to data perturbations together with the associated stability constants 3.(19) for Gaussian elimination and the Doolittle method both with and without partial pivoting. The stability constants of the Doolittle method in high-accuracy arithmetic are around 1/3 of those of Gaussian elimination in standard-floating point arithmetic.

Example 2 comes from the automotive industry and was obtained by a boundary element method. The linear system has a full matrix of $n = 86$ rows and columns which shows no obvious special properties like symmetry, diagonal dominance or the like. The approximate solutions $\bar{x}_i$ were calculated in a floating-point arithmetic (MD 28) using partial pivoting and monotone rounding downwards to 28 binary digits. Table 2 collects some typical data condition numbers, stability constants and solutions x_i together with the error and residual percentages of the approximate solutions. In this example, the error and residual stability constants of the Doolittle method in high-accuracy arithmetic (MD 28) are approximately between 1/25 and 1/30 of those for Gaussian elimination in floating-point arithmetic (MD 28). The error and residual percentages show that our rounding condition numbers $\rho_i^R = \omega_i\rho_i^D$, $\tau_i^R = \psi_i\tau_i^D$ yield realistic measures of the possible errors and residuals.

The third example is a finite element approximation (Bogner-Fox-Schmit element) of a long thin cantilever subdivided into 32 rectangular finite elements as shown in Fig. 1. At each vertex P of the rectangles the function value $v(P)$ of the finite element solution v, and the partial derivatives $hv_x(P)$, $hv_y(P)$, $h^2v_{xy}(P)$ are computed. For example,

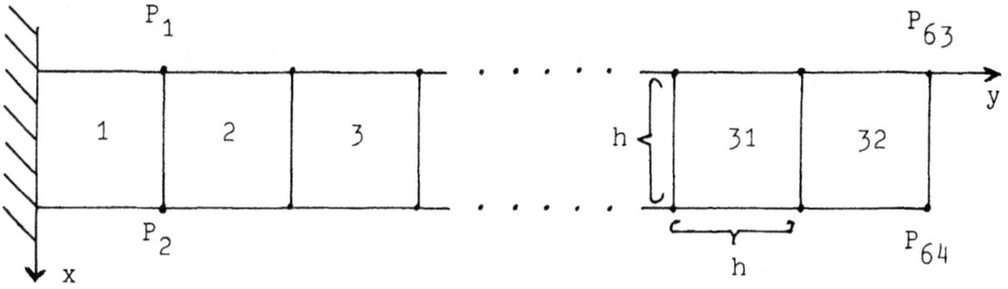

Fig. 1: Thin long cantilever with 32 equal elements

$$x_1 = v(P_1),\ x_2 = hv_x(P_1),\ x_3 = hv_y(P_1),\ x_4 = h^2v_{xy}(P_1),$$
$$\cdots$$
$$x_{253} = v(P_{64}),\ x_{254} = hv_x(P_{64}),\ x_{255} = hv_y(P_{64}),\ x_{256} = h^2v_{xy}(P_{64}).$$

The biggest error is found in the unknown $\bar{x}_{254}$ at the tip of the cantilever. The associated relative rounding condition number for the method of Doolittle is $\bar{\rho}_{254} = 9.80{+}10$ and for Gaussian elimination $\bar{\rho}_{254} = 1.03{+}12$. This shows that maximally up to 11 or 12 decimal digits could be inaccurate in the computed unknown. It turns out that the actual errors of $\bar{x}_{254}$ attain 14 % and 15 %, respectively, of our first order bound $\rho^R_{254}\eta_R$ for the floating-point arithmetic (MD42), using 42 binary digits and monotone rounding downwards. Note that the unknowns $\bar{x}_i$, computed by the method of Doolittle, are only around one decimal digit more accurate than those computed by Gaussian elimination. The error percentages P_i% frequently attain 30 % and more in this example. This is due to the fact that monotone rounding was used and that the matrix A of the linear system has a special structure. In particular, A^{-1} is to a significant extent nonnegative.

Table 3. Finite element approximation of a long thin cantilever, data condition numbers, stability constants, solutions x_i, associated error and residual percentages

		Gauss		Doolittle		
i	ρ_i^D	ω_i	P_i%	ω_i	P_i%	x_i
1	8.02+06	6.98	-30	.764	-29	2.45+02
2	8.44+06	7.16	-30	.757	-30	7.58+01
3	8.03+06	6.99	-30	.763	-29	5.22+02
4	9.48+06	6.94	-14	.808	-13	-1.61+01
253	9.02+06	6.98	-30	.763	-30	1.43+05
254	1.35+11	7.63	14	.726	15	-1.80-02
255	9.32+06	6.98	-30	.762	-30	5.97+03
256	1.20+09	7.77	-24	.686	-38	2.32-02

		Gauss		Doolittle	
i	τ_i^D	ψ_i	Q_i%	ψ_i	Q_i%
1	5.92+07	2.45	- 4	.307	-38
2	1.60+07	1.57	0	.097	34
3	1.97+07	2.62	-18	.321	- 9
4	3.49+06	5.30	-11	.315	15
253	1.07+10	11.1	23	.828	47
254	3.13+09	12.7	6	1.13	-10
255	3.12+09	12.4	4	1.20	- 3
256	6.39+08	13.8	26	1.38	43

References

1. Bowdler, H.J., et al.: Solution of real and complex systems of linear equations. Numer. Math. 8, 217-234 (1966).

2. Fox, L.: An introduction to numerical linear algebra. Oxford: Clarendon Press 1964.

3. Olver, F.W.J., and Wilkinson, J.H.: A posteriori error bounds for Gaussian elimination. IMA J. Num. Analysis 2, 377-406 (1982).

4. Stummel, F.: Optimal error estimates for Gaussian elimination in floating-point arithmetic. Z. Angew. Math. Mech. 62, T 355 - T 357 (1982).

5. Stummel, F.: Forward error analysis of Gaussian elimination. Part I: Error and residual estimates. Numer. Math. (1985). Part II: Stability theorems. Numer. Math. (1985).

6. Stummel, F.: Strict optimal error estimates for Gaussian elimination. Z. Angew. Math. Mech. 65, T 396 - T 398 (1985).

7. Stummel, F.: Strict optimal a posteriori error and residual bounds for Gaussian elimination in floating-point arithmetic. Submitted to Computing.

8. Stummel, F.: FORTRAN-Programs for the rounding error analysis of Gaussian elimination. Centre for Mathematical Analysis, The Australian National University, Canberra, CMA-R02-85.

9. Wilkinson, J.H.: Rounding errors in algebraic processes. Englewood Cliffs: Prentice Hall 1963.

Solving Large Sparse Linear Systems
With Guaranteed Accuracy

U. Schauer

and

R. A. Toupin

1. Introduction

Mathematical models of physical systems are often formulated as integral- differential systems of equations. The construction of approximations to the solution of such systems by any of the standard methods of finite elements or finite differences poses generally the problem of solving a sequence of one or more large systems of linear equations.

$$Ax = b, \tag{1.1}$$

where $x, b \in R^n$, and A is a linear transformation[1]

$$A{:}R^n \to R^n. \tag{1.2}$$

The dimension, or number of unknowns n, may be as small as two, or run into the millions. The matrix representation of the linear operator

$$A = (a_{i,j}) \tag{1.3}$$

is typically, but not always sparse; i.e., of the n^2 elements $a_{i,j}$ only a small percentage are different from zero. For example, the boundary value problem

1 In the following we will also use the notation $x, b \in VR, A \in MR$.

$$\nabla^2\phi(x) = 0, \qquad x \in D \subset R^n$$
$$\phi(x) = h(x), \qquad x \in \partial D \tag{1.4}$$

as normally discretized leads to a system of linear equations in which a small number, say $m = 3, 5$ or 7, of the elements in each row of A are different from zero. Thus the fraction of non zero elements of A, i.e. the ratio (m / n), tends to zero as the dimension of the problem becomes ever greater. In other cases, it may be that the matrix representation of the linear operator A is full, yet a fast and efficient evaluation of Ax is possible without ever computing and storing the matrix elements of A, nor computing the image Ax by the customary matrix-vector inner product. Thus we shall consider here a class of problems, more general than the "sparse matrix" problems, for which the computation required for an evaluation of Ax grows much more slowly than n^2, say like n or $n \log n$. The conjugate gradient method is an appropriate algorithm for solving such problems.

We further narrow our considerations here to the construction and ***verification*** of floating-point approximations to solutions of large linear systems. By this we mean that we select a floating-point system of real numbers $F = F(b, p, e_l, e_u)$ with base b, precision p and exponent range e_l to e_u and compute an ***interval vector*** (n-dimensional parallelepiped)

$$[x] \equiv \{x_i\colon\ l_i \le x_i \le u_i\} \subset R^n,$$

where the lower and upper bounds l_i, u_i are elements of the floating-point system F.[2] A solution of the given system of linear equations is "guaranteed" to lie somewhere within the interval vector $[x]$.

The interval $[x]$ is said to be ***optimal*** if there exists no smaller interval vector properly contained in $[x]$ with these properties.

In general, the elements of the matrix A and the vector b need not be elements of the floating-point system F. Also, if (1.1) arises in modelling a physical system, it may be that some of these real numbers are not known precisely, but only that they lie within some interval of real numbers. It might also be that these coefficients are intervals resulting from some prior computation. Thus a natural generalization of the problem

---

2 For more information on floating-point systems see appendix.

(1.1) is

$$[A]x = [b], \tag{1.5}$$

where $[A]$ is an interval matrix and $[b]$ is an interval vector. The problem then becomes the construction of an interval vector $[x]$, such that the set $\{x\}$ of all solutions of the set of linear equations (1.5) is contained in $[x]$.

We investigate the verification problem alongside two well known algorithms for solving linear systems of equations:

- Sparse Gaussian Elimination methods as developed by Gustavson [*GUS72*]
- Conjugate Gradient Method (CGM) of Hestenes and Stiefel [*HES84*]

Both methods, Gaussian elimination and CGM, admit the use of straight-forward interval arithmetic to obtain verified results. However, it is also well known that useful bounds can be obtained only in rare cases [*NIC77*]. Therefore our solution proceeds differently in two steps :

- Compute a floating-point approximation $\tilde{x}$ of the solution x^* of the problem (1.1) or (1.5)
- Compute lower and upper bounds on the difference $\tilde{x} - x^*$

The existence and the uniqueness of the solution x^* may be known ***a priori***, or may result as part of the second ***verification step***.

The first step can usually be done without interval arithmetic. The second step typically uses interval arithmetic in combination with a fixed point theorem applied to a suitable contracting map.

All results presented have been tested extensively for feasibility and applicability taking into consideration both time and storage requirements. Some of the results were obtained using variable precision arithmetic.

2. *Verification*

Applying Brouwer's fixed point theorem to the map

$$\begin{aligned} f(x) &= x + R(b - Ax) \\ &= Rb + \{I - RA\}x , \end{aligned} \tag{2.1}$$

Rump devised a verification algorithm based on the following theorem [*RUM84*]:

Theorem 2.1 Let $A; R \in MR, b \in VR$. If then for some interval vector $[x]$ and its interior $[\mathring{x}]$,

$$Rb + \{I - RA\}[x] \subseteq [\mathring{x}], \tag{2.2}$$

then R and A are non-singular and there is one and only one $x^* \in [x]$ with $Ax^* = b$.

For large sparse systems of linear equations, time and storage constraints prohibit the use of an approximate inverse R. However, an approximate triangular factorization LU of A can be used with the map, [*HAH83*][3]

$$\begin{aligned} g(x) &= x + (LU)^{-1}(b - Ax) \\ &= (LU)^{-1}(b + \{LU - A\}x) . \end{aligned} \tag{2.3}$$

If for some interval vector $[x]$,

$$g([x]) \subseteq [x], \tag{2.4}$$

then there exists a fixed point for every matrix in $C = [LU - A]$. Especially for $C^* = LU - A$ there exists $x^* \in [x]$ with $Ax^* = b$. Existence of x^* is guaranteed by (2.4), since if L, U exist, then they are non-singular. Uniqueness holds if $g([x]) \subset [x]$, which implies that A is non-singular.

3. *LU-Factorization*

We assume that A is non-singular and that its rows are permuted so that all diagonal pivot elements are nonzero.[4] Choosing L as unit lower triangular matrix, the following formulas for calculation of L and U

---

3 The proposal in the reference to use L^{-1} and U^{-1} in order to obtain closer bounds than by forward/ backward substitution is not attractive for large sparse systems of linear equations.

$$\begin{cases} u_{i,j} = 0 & for\, j < i \\ u_{i,j} = a_{i,j} - \sum_{k=1}^{j-1} l_{i,k} u_{k,j} & for\, j \geq i \end{cases} \tag{3.1}$$

$$\begin{cases} l_{i,j} = 0 & for\ i < j \\ l_{i,j} = 1 & for\ i = j \\ l_{i,j} = (a_{i,j}) - \sum_{k=1}^{n} l_{j,k} u_{k,j} & for\ i > j \end{cases} \tag{3.2}$$

are easily obtained from

$$a_{i,j} = \sum_{k=1}^{n} l_{i,k} u_{k,i} \quad for \quad i,j = 1, \dots n. \tag{3.3}$$

If the computations (3.1) and (3.2) are carried out in the usual way using some floating-point system $F(b, p, e_l, e_u)$ and rounding strategy, they yield only some floating-point approximation $\tilde{l}_{i,j}\, \tilde{u}_{i,j}$ for $l_{i,j}\ u_{i,j}$. With high enough precision[5] the differences between $\tilde{L}$, $\tilde{U}$ and L, U can be made arbitrarily small : for every $\varepsilon > 0$ there is a precision $p \geq p(\varepsilon)$ such that

$$\begin{aligned} \| \tilde{L}\tilde{U} - A \|_\infty &= \max_i \sum_{j=1}^{n} | \tilde{L}\tilde{U} - A |_{i,j} \\ &< \varepsilon \max_i \sum_{j=1}^{n} | a_{i,j} | \\ &= \varepsilon \| A \|_\infty . \end{aligned} \tag{3.4}$$

Assume

$$\varepsilon \| A \|_\infty \ \| A^{-1} \|_\infty \leq q < \frac{1}{2} . \tag{3.5}$$

---

4 Pivoting for size within rows or columns could be easily combined with the calculation process.

5 Accurate evaluation of scalar products (see appendix) may help to bound the required precision.

Then

$$\| \tilde{L}\tilde{U} - A \|_\infty \| A^{-1} \|_\infty \le q < \frac{1}{2} ,$$

and

$$\| (\tilde{L}\tilde{U})^{-1} \|_\infty \le \frac{\| A^{-1} \|_\infty}{1 - q} .$$

Therefore,

$$q \ge \| \tilde{L}\tilde{U} - A \|_\infty \| A^{-1} \|_\infty$$

$$\ge (1 - q) \| \tilde{L}\tilde{U} - A \|_\infty \| (\tilde{L}\tilde{U})^{-1} \|_\infty ,$$

and

$$\| A^{-1} \|_\infty \le \frac{1 - q}{1 - 2q} \| (\tilde{L}\tilde{U})^{-1} \|_\infty .$$

As a consequence of (3.5) we obtain for an approximate solution $\tilde{x}$ with $\tilde{r} = A\tilde{x} - b$ the ***a posteriori*** error estimate

$$\| x^* - \tilde{x} \|_\infty \tag{3.6}$$

$$\le \| \tilde{r} \|_\infty \| A^{-1} \|_\infty$$

$$\le \frac{1 - q}{1 - 2q} \| \tilde{r} \|_\infty \| (\tilde{L}\tilde{U})^{-1} \|_\infty .$$

Lemma 3.1

For any problem (1.1) with non singular matrix A one can obtain a verified solution by means of triangular factorization provided the precision of the floating point system F is sufficiently high.

From the calculation process (3.1), (3.2) the following properties of the LU-factorization[6] are quite obvious.

---

6 In the following LU will always be used in the sense of $\tilde{L}\tilde{U}$.

- Using high precision scalar products (see appendix), as provided in ACRITH [*IBM83*] the interval matrix $C = [LU - A]$ can be computed accurately (with only one rounding operation) and efficiently[7] as a byproduct of the LU-factorization.

- By means of properly chosen directed roundings for the elements of L and U, one can always achieve either $C \geq 0$, or $C \leq 0$.

- C has the same sparseness as the triangular factors L and U.

A bound for $\| (LU)^{-1} \|_\infty$ can be obtained by a forward/backward substitution process applied to the unit cube. As pointed out by Nickel [*NIC77*] we must expect an overestimation of the order 2^n. The precision of the floating-point system $F(16, 14, -64, 63)$ is sufficiently high only for Hilbert matrices [8] of dimensions not exceeding 9, and the precision of $F(16, 28, -64, 63)$ for dimensions up to 18.

Verification based upon LU- factorization (i.e.,2.4) is, in general, 6 times faster than verification based upon an approximate inverse (i.e.,2.2). Assuming that the time complexity of multiplication grows quadratically with the precision, then doubling the precision and using (2.4), is still faster than (2.2).

Our experience with sparse systems of linear equations, though limited, seems to confirm the effectiveness of verification based on LU-factorization. The overestimation of the forward/backward substitution process is generally much lower than 2^n for sparse matrices. Using the floating-point system $F(16, 14, -64, 63)$ the solution of sparse matrices with up to one thousand unknowns could be verified successfully if the average number of nonzero coefficients per row were ≤ 10.

4. Contraction

The map (2.1) contracts the parallelepiped

$$P = \{x: |x - \tilde{x}|_i \leq \alpha |\tilde{x}_i|\} \tag{4.1}$$

---

7 At the cost of only one addition per element of C plus a multiplication for the subdigonal elements.

8 By Hilbert matrices we mean (h_{ij}) with $h_{ij} = \frac{1}{i+j-1}$.

if and only if every vertex $\tilde{x}^{(k)}$, $k = 1, \ldots, 2^n$ of P maps into the interior of P. That is, if and only if

$$|f(\tilde{x}^{(k)}) - \tilde{x}|_i < \alpha|\tilde{x}|_i \begin{cases} i = & 1, \ldots, n \\ k = & 1, \ldots, 2^n \end{cases} \quad (4.2)$$

Lemma 4.1

Sufficient conditions that the map (2.1) contracts P are:

$$\tilde{x} \neq 0, \tag{4.3}$$

$$|\tilde{x}|_i > \sum_{j=l}^{n} |I - RA|_{i,j}|\tilde{x}|_j, \tag{4.4}$$

and

$$\alpha > \max_i \frac{|R(b - A\tilde{x})|_i}{|\tilde{x}|_i - \sum_{j=1}^{n} |I - RA|_{i,j}|\tilde{x}|_j} \; . \tag{4.5}$$

The proof is straight forward; (4.5) is obtained by looking at the worst possible vertex.

Suppose $\tilde{x}$ approximates a solution x^* with one or more components $x^* = 0$. Then one or more of the components $\tilde{x}$ might well be 0, and Lemma 4.1 does not apply. In such a case, we consider in place of the linear system $Ax = b$ the system

$$Ay = \tilde{b}, \;\; \tilde{b} = b + A(1 - \tilde{x}) \; , \tag{4.6}$$

and choose for the estimate $\tilde{y}$ of a solution to this system, $\tilde{y} = 1$. With this choice for $\tilde{y}$ the residuals are equal:

$$b - Ax = \tilde{b} - A\tilde{y}. \tag{4.7}$$

Theorem 4.2

If for $i = 1, 2, \ldots n$

$$\sum_{j=1}^{1} |I - RA|_{ij} \;\; < 1$$

and

$$\alpha > \frac{\max_i |R(b - A\tilde{x})|_i}{(1 - \sum_{j=1}^{n} |I - RA|_{i,j})},$$

then (2.1) has a unique fixed point x^* with $b - Ax^* = 0$ and $|x^* - \tilde{x}|_i < \alpha$.

Proof: By lemma 4.1 the map $f(y) = y + R(\tilde{b} - A\tilde{y})$ is contracting and by Brouwer's fixed point theorem there exists a fixed point y^* with $\tilde{b} - Ay^* = 0$. y^* is uniquely determined since the mapping is into the interior and $x^* = y^* + \tilde{x} - 1$ is a solution of $b - Ax^* = 0$. That $|x^* - \tilde{x}|_i < \alpha$ follows immediately from $|y^* - \tilde{y}|_i < \alpha$ and the definition (4.6).

We use Theorem (4.2) with $R = (LU)^{-1}$ to prove the existence and uniqueness of the solution x^* of $b - Ax = 0$ as follows: First one has to determine if

$$\sum_{j=1}^{n} |(LU)^{-1}(LU - A)|_{i,j} < 1 \tag{4.8}$$

- Bounds for the left hand side of (4.8) can be obtained by applying a forward/backward substitution to the interval vector $[-d_i, +d_i]$ with

$$d_i = \sum_{j=1}^{n} \max |LU - A|_{i,j} \,, \tag{4.9}$$

  where the maximum value must be taken for each component of the interval matrix $[LU - A]$.

- The interval matrix $[LU - A]$ is not needed explicitly, only the row sums of the absolutely largest bounds. These can be accumulated efficiently during LU-factorization.

- Condition (4.8) does not depend on the approximation $\tilde{x}$, but only on the LU-factorization. If it holds, then iterative residual correction can be used to improve the estimate $\tilde{x}$ if required.

The crucial point remains the overestimation inherent in the substitution process applied to intervals. Even if (4.8) be true, one may not be able to prove it with interval arithmetic if the precision is not sufficiently high. Nickel [*NIC77*] reports that this difficulty can be avoided for M-matrices.

5. *Verification for M-matrices*

The following definition is from Varga [*VAR62*].

Definition 5.1: A real $n \times n$ matrix $A = (a_{i,j})$ with $a_{i,j} \leq 0$ for all $i \neq j$ is an M-matrix if A is non-singular and $A^{-1} \geq 0$.

M-matrices arise in many applications, especially elliptic boundary value problems. For M-matrices Lemma 4.1 can be improved as follows.

Lemma 5.2: Assume A is an M-matrix,

$$A\tilde{x} > 0 \ or \ A\tilde{x} < 0\,, \tag{5.1}$$

and that

$$\alpha > \max_i \frac{|b - A\tilde{x}|_i}{|A\tilde{x}|_i}, \tag{5.2}$$

then

$$|f(\tilde{x}^{(k)}) - \tilde{x}|_i < \alpha|\tilde{x}|_i\,.$$

Proof: $\tilde{x}_i = (A^{-1}(A\tilde{x}))_i \neq 0$, and $|\tilde{r}| = |b - A\tilde{x}| < \alpha|A\tilde{x}|$ imply that

$$|A^{-1}\tilde{r}| \leq A^{-1}|\tilde{r}| < \alpha A^{-1}\,|A\tilde{x}| = \alpha|\tilde{x}|$$

so that Lemma (4.1) applies with $R = A^{-1}$.

If neither $b > 0$ nor $b < 0$ then the assumption (5.1) will not hold true, if $\tilde{x}$ approximates the solution x^*. However, using the transformations

$$\begin{aligned} y(x) &= x + \bar{x}, \quad A\tilde{y} = A(\tilde{x} + \bar{x}) \\ \bar{b} &= b + A\bar{x} \\ \tilde{r} &= b - A\tilde{x} \quad = \bar{b} - A\tilde{y} \end{aligned} \tag{5.3}$$

and applying Lemma (5.2) to $\bar{b} - Ay = 0$ we obtain

Theorem 5.3 If A is an M-matrix, and either $A(\tilde{x} + \bar{x}) > 0$, or $A(\tilde{x} + \bar{x}) < 0$, and

$$\alpha > \max \frac{|b - A\tilde{x}|_i}{|A(\tilde{x} + \bar{x})|_i},$$

then $|x^* - \tilde{x}|_i < \alpha|\tilde{x} + \bar{x}|_i$, where $b - Ax^* = 0$.

A suitable $\bar{x}$ giving $A(\tilde{x} + \bar{x})$ may be obtained as an approximate solution of $A\tilde{y} = 1$, i.e. $A\bar{x} = 1 - A\tilde{x}$, or, if $\tilde{x}$ is a close approximation to x^* as an approximate solution of $A\bar{x} = 1 - b$. The condition that A be an M-matrix can be weakened. To prove lemma (5.2) we need only $A^{-1} \geq 0$.[9] Therefore theorem (5.3) applies to matrices obtained if Hilbert matrices are multiplied from left and right with the diagonal matrix $(-\delta_{ij}(-1)^i)$. These matrices are monotone but not M-matrices and, like the Hilbert matrices, are ill conditioned. Experiments with the transformed Hilbert matrices show that one must use sufficiently high precision arithmetic to exhibit a positive lower bound for $A(\tilde{x} + \bar{x})$. The precision of the floating point system $F(16, 14, -64, 63)$ is sufficiently high only for the transformed Hilbert matrices of dimensions ≤ 12. If $A(\tilde{x} + \bar{x})$ is evaluated differently as
$(A, A)\begin{pmatrix} \tilde{x} \\ \bar{x} \end{pmatrix}$ [10] then verification can be obtained for dimensions ≤ 16. For even higher dimension the calculation of the approximations $\tilde{x}$ and $\bar{x}$ needs higher precision (for both LU factorization and forward/backward substitution).

---

9 Non-singular matrices A with non-negative inverse represent monotone operators [*COL64*]: $Ax \geq Ay$ implies $x \geq y$.

10 To avoid the truncation error inherent in forming $\tilde{x} + \bar{x}$.

6. Conjugate Gradient Method (CGM)

For a typical application of CGM we expect the number of unknowns n to be very large, say of the order of 1 000 to 100 000, and we hope that the number of iterations for obtaining sufficient accuracy is significantly less than n, say of the order $\sqrt{n}$.

For M-matrices (elliptic boundary value problems) lemma 5.2 may serve as a stopping rule (if $A\tilde{x} > 0$ or $A\tilde{x} < 0$): Iteration is stopped, when $\alpha > \max|\tilde{r}_i| / |A\tilde{x}|_i$ is small enough, guaranteeing $|x^* - \tilde{x}|_i < \alpha|\tilde{x}|_i$

Experiment 1 To study the properties of CGM we applied it in variable precision to the monotone transformed Hilbert matrices. The right-hand side vector was chosen in compliance with a vector of all ones for the unmodified Hilbert matrices.

An overview of the results is displayed in the table[11] below with condition, precision, and accuracy each measured in number of decimal digits.

11 If several values appear in the precision and accuracy columns, then the first, second,... values belong together.

Dimension	*Condition*	*Precision* [12]	*Accuracy* [13]
5	6	26	15
6	8	26,34	8,19
7	9	34 45	10,20
8	11	45,55	12,21
9	12	55.65	10,20
10	14	65,74	8,18
11	15	74,84	6,16
12	17	84,93,103	2,12,21
13	18	103,113	7,17
14	20	122,132	10,20
15	21	142,151,161	11,21,31
16	23	170,180,190	12,22,31
17	24	180,190,199	2,11,21
18	26	190,199,209	1,10,20
19	27	219,228	7,17
20	29	238,248,	4,13
21	30	267,276,286, 296	7,17,27,36

Observations :

There exists a **Critical Precision P(A)** for the CGM such that:

- If the precison of the arithmetic is greater than P(A) the residual diminishes monotonically to zero in a finite number of steps less than or equal the dimension of A.

- For high enough precision the accuracy of the approximate solution increases at the same rate as the precision.

---

12 The granularity of the multiple precision package was 2^{32} (8 Hex digits).

13 Theorem 5.3 was used to determine the number of accurate digits.

- Insufficient precision can be detected dynamically by monotoring the residuals.

- An adequate precision can be found systematically with a worst case penalty factor of 4 by multiplying the precision at each successive trial with $\sqrt{2}$. [14]

- As a stopping rule for the general case one may use "smallness" of the residuals. In case of M-matrices verified results are obtained by exploiting Lemma 5.2 or Theorem 5.3.

- The time needed for verification is negligible if the right-hand side vector b is all positive or all negative, otherwise it may at worst equal the time needed for calculation of the approximation $\tilde{x}$. [15]

Experiment 2: As a model elliptic boundary value problem we chose the Laplace equation over a rectangle with various types of boundaries (considering only meshpoints) and boundary conditions of Dirichlet and mixed Dirichlet/Neumann type.

The results obtained for the Dirichlet-problem over an m by m square with the boundary values shown below were as follows :

$$\begin{array}{cl} 1 & 2\ 3\ \ldots\ m \\ m+1 & .\ .\ \ldots\ 2m \\ . & .\ .\ \ldots\ . \\ (m+1)m+1 & .\ .\ \ldots\ mm \end{array}$$

- The number of iterations was $O(m)$ with a factor in the range 0.75...4.55, for $m = 4, ..., 44$, to obtain a relative accuracy of the order $3 \cdot 10^{-14}$. For $m = 100$ the relative accuracy obtained after 431 iterations was $\sim 1.78 \cdot 10^{-13}$.

- In the indicated range, calculation could be performed safely in the floating-point system $F(16, 14, -64, 63)$. This may be largely due to the simple coefficients 1 and -0.25 of the finite difference approximation and to some extent also to the uniform order of magnitude of the boundary values.

---

14 This is a consequence of the quadratic law of time dependency on precision.

15 If CGM with adjusted precision is used to calculate $\bar{x}$ as solution of $A\bar{x} = b - A\tilde{x}$.

- The heuristic stopping rule, to terminate when the value of the residuals had decreased by a factor of 10^{-16}, provided the above relative accuracy.

- For verification of the obtained accuracy the additional effort never exceeded 20% of the time spent to calculate the approximation.

Verification was based on theorem 5.3. The right-hand side of our equation system obviously has many zeros; therefore, the calculation of $\bar{x}$, solving $A\bar{x} = 1 - A\tilde{x}$ at least approximately, such that $A(\tilde{x} + \bar{x}) > 0$ is necessary.

Our problem can be written as $Ax \equiv (I - N)x = b$ with $\| N \| < 1$, since A is irreducibly diagonal dominant. Therefore a finite part of the Neumann series provides a good approximation to A^{-1} and the computational effort can be finely adjusted to the requirement $A(\tilde{x} + \bar{x}) > 0$.

The verification based upon theorem 2.1 with a finite part of the Neumann series as approximate inverse turned out to be less efficient. However, it can still be applied if A is not an M-matrix, but only irreducibly diagonal dominant.

Our experiment suggests that CGM with high enough precision may yield an approximate solution of some desired accuracy in fewer steps and less overall time than any alternative, in spite of the higher cost for extra precision arithmetic. On the other hand, in contexts where many linear systems with the same or similar sparsity structure must be solved, SGE may be the preferred method. An investigation of the "adjustment problem" for large sparse geodetic networks [*SCH77*] showed that SGE may perform much better (for such non-linear problems) than CGM (the observed factor CGM/SGE was of the order 5 to 10).

Summary :

Further investigations are desirable in several directions :

- The discretized equation system of the Laplace equation was surprisingly well - conditioned. Additional experiments with boundary value problems leading to ill - conditioned equation systems may be needed to asses the importance of variable precision for CGM.

- The conjugate gradient method can be speeded up by different means, e.g. incomplete *LU* decomposition [*MEI77*] and polynomial preconditioners [*JOH82*]. It seems that CGM combined with such techniques has the potential to "solve" large problems efficiently.

- The verification technique based on lemma 5.2 or theorem 5.3 may be applicable for even wider problem classes than M-matrices or monotone matrices. It seems to qualify as a heuristic stopping rule which is superior to "smallness of the residuals".

Appendix: Summation of Floating Point Numbers.

Most implementations of floating-point arithmetic deliver as a result for the sum of the three numbers

$$10^{50}, 10^{30}, -10^{50} \tag{A.1}$$

the value 0 or 10^{30}, depending on the ordering of the summands. Our objective in this appendix is to analyse thoroughly the algorithm which behaves in this seemingly peculiar way. The algorithm is well known, and analyses of it have previously appeared in textbooks and journals. Some of these prior analyses are incomplete, and some reach demonstrably incorrect conclusions.

Let $b \geq 2$, $p \geq 1$, $-e_l \leq 0 \leq e_u$ denote integers subject to the restrictions noted. Every such choice of these parameters serves to define a finite set of rational numbers $F = F(b, p, e_l, e_u)$. called a floating-point system.

The floating-point system F comprises the set of rational numbers

$$F \equiv \{x = \pm mb^e: 0 \leq m < 1, e_l \leq e \leq e_u\}, \tag{A.2}$$

The integers b, p, e_l, e_u are called, respectively, the base, precision, smallest exponent, and largest exponent of the the system. If $x = \pm mb^e$ is an element of F then m is called the *mantissa* of x, and e is its exponent. In general, a given element of F does not have a unique mantissa and exponent. Uniqueness is provided by the rules

$$m \geq b^{-1} \textit{ for } e > e_l$$
$$m = 0 \textit{ for } e = 0$$

Some digital hardware and floating-point software is designed to treat only normalized floating-point systems. These differ in a seemingly innocuous way from a floating-point system as just defined.

A normalized floating-point system is a floating-point system as defined above with the elements having mantissae not in the range $0 < m < b^{-1}$. This excludes those elements of F having the smallest exponent and a mantissa with leading digit equal to zero in its binary representation.[16] Consider the intervals $[a, b]$ defined by successive elements a, b of a floating-point system F. The real line is the union of these closed intervals and the pair of unbounded intervals $[\infty, -M]$, $[M, \infty]$, where $M = b^{e_u}(1 - b^{-p})$ is the maximum element of F. Let I denote the set of all these intervals which cover the real line. The pair of these intervals containing the origin and the unbounded pair we call irregular, the others are called regular. The intervals of smallest width $|a - b|$ lie adjacent to the origin 0 and, proceeding to the left or right of the origin, the width of successive intervals does not diminish. In a normalized floating-point system, the intervals of smallest width are not adjacent to the origin. Rather, they are preceeded to the left or right of the origin by a single interval of greater width.

If $\tilde{x}$ is an endpoint of a regular interval $i \in I$ then it approximates any $x \in i$ with relative error $0 \leq \varepsilon \leq \varepsilon^* \equiv b^{-p}$. That is to say, there exists an $0 \leq \varepsilon \leq \varepsilon^*$ such that

$$x = (1 \pm \varepsilon)\,\tilde{x}, \quad x \in i, \quad \tilde{x} \in F \tag{A.3}$$

or, equivalently,

$$|x - \tilde{x}| \leq \varepsilon^* \,|\tilde{x}|, \quad \tilde{x} \in F, \quad \forall x \in i. \tag{A.4}$$

No such equalities or inequalities hold for irregular intervals. Either endpoint of a regular interval approximates all its points with a uniform relative error ε^*, which can be made as small as we please by suitable choice of the precision of the system. For irregular intervals containing the origin, we can write

$$|x - \tilde{x}| \leq \varepsilon \,|\tilde{x}|, \quad \tilde{x} = b^{e_l - p} ,$$

but, the relative error ε is unbounded, independent of the precision.

The sum $x + y$ of two floating-point numbers $x, y \in F$ is not, in general, a member of the same floating-point system. However, the sum never falls in the interior of an irregular interval containing the origin (underflow). It may of course lie in one of the unbounded intervals (overflow). Thus if $x, y \in F$, the sum $x + y$ can always be approx-

---

.6 Kulisch and Miranker [*KUL80*] call a floating-point system as we have defined it above, an extended floating-point system. What we have called a normalized system, they call a "floating-point system". Obviously, their terminology is suggested by regarding the normalized systems as the primary object and of adding elements or "extending" these sets to obtain what we call a floating-point system. Our terminology and viewpoint lead us to call a normalized floating-point system "defective", meaning that it is obtained from a floating-point system by deletion of certain of its elements.

imated in accordance with (A.3-4) by an element of F provided $|x+y| \leq M$. This is not true for normalized floating-point systems because underflow may occur.

Suppose one has a "machine" or "subroutine" that takes as input any pair of elements $\{x, y\}$ of a floating-point system F and delivers as output an element $x \tilde{+} y \in F$ such that

$$x \tilde{+} y = (1 \pm \varepsilon)(x+y), \quad 0 \leq \varepsilon \leq \varepsilon^* \qquad \text{(A.5)}$$

whenever such a element exists. It will exist for all pairs in F except those for which $|x+y| > M$. We suppose that in those cases the machine delivers the result x_∞ (signals the overflow condition).

Consider then the problem of computing an approximation $\tilde{S} \in F$ to the sum

$$S = \sum_{k=1}^{n\geq 3} x_k \qquad \text{(A.6)}$$

of three or more floating-point numbers. The most commonly used algorithm for this purpose is to compute $\tilde{S}$ according to the scheme

$$\tilde{S} = (((x_1 \tilde{+} x_2) \tilde{+} x_3) \tilde{+} x_4)... \qquad \text{(A.7)}$$

In general, there exists no $\varepsilon \leq \varepsilon^*$ such that

$$S = (1 \pm \varepsilon)\tilde{S} \qquad \text{(A.8)}$$

or, equivalently, no ε in this range such that

$$|S - \tilde{S}| \leq \varepsilon|\tilde{S}| \qquad \text{(A.9)}$$

If $F = F(16, 14, -64, 63)$ then example (A.1) suffices to prove our last assertion. Note that the result delivered by the algorithm (A.7) depends on the ordering of the summands.

In place of (A.9) we can show that, provided no overflow occurs, for all vectors $x = (x_i)$ with $x_i \in F$

$$|S - \widetilde{S}| \leq \varepsilon^* \| \iota_n \|_2 \| x \|_2 , \tag{A.10}$$

where $\| \ \|_2$ denotes the Euclidean norm and ι_n is the vector with integer components 1, 2, 3, ...n.

Proof of (A.10): From the definition (A.7) of $\widetilde{S}$ and the property (A.5) of $\widetilde{+}$ we first conclude that

$$\widetilde{S} = \sum_{k=1}^{n} q_k x_k , \tag{A.11}$$

where the positive numbers q_k are of the form

$$q_k = \prod_{k}^{n} (1 \pm \varepsilon_k) \tag{A.12}$$

and

$$\varepsilon_1 = 0, \ \ 0 \leq \varepsilon_k \leq \varepsilon^* . \tag{A.13}$$

It follows that

$$\begin{aligned} q_k &\geq \prod_{k=1}^{n} (1 - \varepsilon_k) \\ &\geq \prod_{k=1}^{n} (1 - \varepsilon^*) \\ &= (1 - \varepsilon^*)^{n-k+1} \\ &\geq 1 - (n - k + 1)\varepsilon^* . \end{aligned} \tag{A.14}$$

Thus

$$|S - \widetilde{S}| \leq \varepsilon^* \sum_{k=1}^{n} x_k (n - k + 1) . \tag{A.15}$$

It follows from the Cauchy-Schwarz inequality that

$$|S - \tilde{S}| \leq \varepsilon^* \parallel \iota_n \parallel_2 \parallel x \parallel_2 , \tag{A.16}$$

which was to be proved. Note that $\parallel \iota_n \parallel_2 = O(n^3 / 2)$.

If the addends x_k are all of the same sign, then $x_j x_k \geq 0$, and

$$S^2 = \sum_{k=1}^{n} (x_k)^2 + \sum_{j \neq k} x_j x_k \geq \parallel x \parallel_2^2 .$$

In this case, inequality (A.15) implies the inequalities

$$\begin{aligned} |S - \tilde{S}| &\leq \varepsilon^* \parallel \iota_n \parallel_2 |S| \\ &\leq \varepsilon^* \parallel \iota \parallel_2 |S - \tilde{S} + \tilde{S}| \\ &\leq \varepsilon^* \parallel \iota_n \parallel_2 (|S - \tilde{S}| + |\tilde{S}|) . \end{aligned}$$

Thus provided that

$$1 - \varepsilon^* \parallel \iota_n \parallel_2 > 0 \tag{A.17}$$

one sees that

$$|S - \tilde{S}| \leq \frac{\varepsilon^* \parallel \iota_n \parallel_2}{1 - \varepsilon^* \parallel \iota_n \parallel_2} | \tilde{S} | . \tag{A.18}$$

Inequality (A.18) shows that the commonly used algorithm (A.7) for computing an approximation to the sum of three or more floating-point numbers produces a result $\tilde{S}$ with relative error guaranteed to be less than

$$\varepsilon = \frac{\varepsilon^* \parallel \iota_n \parallel_2}{1 - \varepsilon^* \parallel \iota_n \parallel_2} \tag{A.19}$$

provided all the addends have the same sign. The bound on the relative error depends on the number of addends. The algorithm (A.7) fails, in general, to produce an approximation satisfying (A.8) even when the addends are all of the same sign. Of course there is some higher precision floating-point system depending on the number of addends which will deliver an $\tilde{S}$ satisfying (A.8). A sufficiently high precision is easily inferred from the formula (A.19).

If the addends are not all of the same sign then algorithm (A.10) can fail miserably in producing an $\tilde{S}$ with small relative error, as shown by our simple example (A.1). It might occur at first glance that using (A.7) to compute the sum of all the positive addends and the sum of all the negative addends and then the sum of these parts would improve the accuracy of the approximation. Again example (A.1) suffices to disprove the generality of this strategy. What can be shown is the following. Let $\tilde{S}_+ and \tilde{S}_-$ denote the approximations to the sum of all the positive addends and the sum of all the negative addends, respectively, as computed by (A.7). According to (A.18-19) we shall then have

$$|S_+ - \tilde{S}_+| \le \varepsilon_n |\tilde{S}_+| \ , \qquad |S_- - \tilde{S}_-| \le \varepsilon_m |\tilde{S}_-| \ , \tag{A.20}$$

where n and m are the number of positive addends and negative addends, respectively. Let

$$S = S_+ + S_-, \quad \tilde{S} = \tilde{S}_+ \tilde{+} \tilde{S}_- \ , \tag{A.21}$$

When $\tilde{S} \ne 0$ one can show that

$$|S - \tilde{S}| \le \varepsilon^* [1 + \frac{\varepsilon_n S_+ - \varepsilon_m \tilde{S}_-}{|\tilde{S}|}] \, |\tilde{S}| \ . \tag{A.23}$$

This inequality provides an easy computation of the maximum relative error for this method of computing the sum of three or more addends not all of the same sign.

Observe that since the x_k are bounded, for every n and every floating-point system, there exists a sufficiently high precision such that all the partial sums (A.7) are computed without error; i.e., $\varepsilon_k = 0, k = 1, 2, ...n$. Thus S can be computed without error and then rounded so as to satisfy (A.9) if no overflow occurs. In essence this is the method of summation used in ACRITH. Various algorithms are known for computing approximations to the sum of three or more floating-point numbers satisfying (A.9) [*BOH77*], [*DEK71*], [*KAH65*]. The relative speed of these alternatives depends, in general, on the hardware and the floating-point system under consideration; e.g., its precision and exponent range.

To compute the inner product

$$x \bullet y = \sum_{k=1}^{n} x_k y_k$$

of a pair of vectors $x, y \in R^n$ it is customary to compute approximations

$$\begin{aligned} z_k &= x_k \tilde{\times} y_k \\ &= (1 \pm \eta_k) x_k \times y_k, \quad 0 \le \eta_k \le \varepsilon^* \end{aligned}$$

to the required products and sum the results using (A.7). This leads to a formula similar to (A.11-12):

$$x \bullet y = \sum_{k=1}^{n} q_k z_k$$

where the q_k are now given by

$$q_k = \prod_{k}^{n} (1 \pm \varepsilon_k)(1 \pm \eta_k)$$

and $0 \le \varepsilon_k, \eta_k \le \varepsilon^* = b^{-p}$. By reasoning similar to that leading to (A.16) we get

$$|x \bullet y - x \tilde{\bullet} y| \le \varepsilon^* \, \| \iota_{n+1} \|_2 \, \| x \times y \|_2 \, .$$

Here $x \times y$ denotes the vector with components $x_k y_k$. Thus we see that the failure of this algorithm to produce an approximation to the inner product with small relative error can be traced to the weakness of the method of summation. The error caused by approximating the products only increases the factor ι_n to ι_{n+1} in passing from (A.16) to (A.23). The square of the Euclidean norm of a vector is a special case of an nner product in which all of the addends are positive. Thus, in this special case, the lgorithm does provide an estimate of the form (A.18) with an ε dependent on $(n + 1)$.

Bibliography

[BOH77] Bohlender, G. (1977), *Genaue Summation von Gleitkommazahlen.* Pp.21-32 in : Computing Supplementum 1 Grundlagen der Computer Arithmetik (Springer, Berlin- Heidelberg- New York).

[COL64] Collatz,L. (1964), *Funktionalanalysis und numerische Mathematik.* (Springer, Berlin- Heidelberg- New York).

[DEK71] Dekker,T.J. (1971), *A floating-point technique for extending the available precision.* Pp.224-242 in : Numerische Mathematik **18** (Springer, Berlin- Heidelberg- New York).

[GUS72] Gustavson, F. G. (1972), *Some basic techniques for solving sparse systems of linear equations.* Pp. 41-52 in : Rose, D.J., Willoughby, R.A. eds. Sparse matrices and their applications. Proceedings of a Symposium 1971 (Plenum Press, New York).

[HAH83] Hahn, W., Mohr, K. Schauer, U. (1983), *Some techniques for solving linear equation systems with guarantee.* (IBM technical note TN 83.01) to appear in Computing.

[HES84] Hestenes, M. R. (1984), *Conjugate direction methods in optimization.* (Springer, Berlin Heidelberg- New York).

[IBM83] *ACRITH High accuracy arithmetic subroutine library. General information manual IBM-GC33 - 6163. Program description and users guide IBM-SC33 - 6164.*

[JOH82] Johnson, O. G., Michelli, C. A., Paul, G.(1982), *Polynomial preconditioners for conjugate gradient calculations.* (IBM technical report RC9208).

[KAH65] Kahan, W.(1965), *Further remarks on reducing truncation errors.* P.40 in : Communications of the ACM Vol.8.

[KUL80] Kulisch, U., Miranker, W. L.(1980), *Computer arithmetic in theory and practice.* (Academic Press, London-New York- San Francisco).

[MEI77] Meijerink, J. A., Vorst, H. A.(1977), *An iterative solution method for linear systems of which the coefficient matrix is a symmetric M-matrix.* Pp. 148-162 in : Mathematics of computation 31,137.

[NIC77] Nickel, K. (1977), *Interval analysis.* Pp. 193-222 in : Jacobs, D. ed The state of the art in numerical analysis, Proceedings of a conference 1976 (Academic Press, London-New York- San Francisco).

[SCH77] Schek, H., Steidler, F., Schauer, U. (1977), *Ausgleichung grosser geodätischer Netze mit Verfahren für schwach besetzte Matrizen.* Theoretische Geodäsie **87** (Bayrische Akademie der Wissenschaften).

[RUM83] Rump, S. M. (1983), *Solving Algebraic Problems with high accuracy.* Pp. 53-118 in : Kulisch, U., Miranker, W.L. eds. A new approach to scientific computation, Proceedings of a symposium 1982. (Academic Press, London-New York- San Francisco).

Symbolic and Numeric Manipulation of Integrals

J. H. Davenport,
School of Mathematics,
University of Bath,
Bath BA2 7AY,
England

Summary. In this paper, we look at various techniques of integration. Computers have obviously revolutionised the theory and practice of numeric integration, but they have equally revolutionised the field of symbolic integration. We describe the current state of symbolic integration, and discuss the impact of this art on numeric integration.

Introduction

Symbolic integration is, historically, the major vehicle of integration as a tool of science. Newton used symbolic integration extensively, while Simpson's rule appeared about eighty years later, in 1743. In the early nineteenth century, Liouville, Chebyshev and their students made major progress in the field of symbolic integration, but the work rapidly lead into the complexities of algebraic geometry, and was ousted from undergraduate courses by the rise of other parts of the calculus. The invention of calculators and computers has made numerical integration an easier, indeed almost painless, task for most scientists and engineers.

But computers have a far greater potential impact on mathematics than just the performance of routine (or non-routine) *numerical* calculations. They can equally well be used for *non-numeric* calculations: a possibility that was realised very early [Lovelace, 1844; Nolan, 1953]. Symbolic manipulation of functions is relatively easy in principle as far as addition, subtraction, multiplication, division and differentiation are concerned, for the methods taught at school are essentially algorithmic (though they are often not the most efficient algorithms, and great intellectual effort has been expended by the practitioners of computer algebra in extending them). Indeed, numeric differentiation has many disadvantages compared with symbolic differentiation, being less stable and more expensive [Rall, 1981]. Integration is different, since the methods taught at school are essentially heuristic rather than algorithmic. While it would be possible to build heuristic programs that emulate human integration techniques, better results have in general been achieved by algorithmic integrators.

All timings reported in this paper were measured on a CII-Honeywell 68/80- DPS3 at the Centre Interuniversitaire de Calcul de Grenoble, running under Multics 9.1. The author is grateful to Y. Siret and the staff of the Centre for their assistance. The times shown (in seconds) are those for the calculations alone (no I/O) and seem to be reliable to within 10%. The programs were written in FORTRAN, and used NAG [1982] routines (coincidentally, also Mark 9) where possible. The names of the NAG routines are given in parentheses. The symbolic integration packages used are those of REDUCE [Norman & Moore, 1977] and MACSYMA [Moses, 1971]. The author is grateful to numerous colleagues for their advice and assistance with the preparationof this paper, an earlier version of which appeared in the proceedings of *Tools Methods and Languages for Scientific and Engineering Computation* (ed. B. Forn, J.-C. Rault & F. Thomasset).

Algorithms for Symbolic Integration

The most-quoted paper on symbolic integration is that of Risch [1969], which marks the re-discovery of symbolic integration as a field to which computers, and algorithmic algebra, can be applied. In order to state his, and more recent, results, we need some terminology.

We say that a function is *algebraic* if it is a root of some polynomial. A trivial example of this is $\sqrt{x}$, which is a root θ of $\theta^2-x = 0$. However, there are algebraic functions which can not be expressed in terms of n-th roots, for example the roots of $\theta^5-\theta-x = 0$. This notation also includes functions expressible as roots of polynomials whose coefficients include other algebraic functions, such as $\sqrt{(1+\sqrt{x})}$, which is a root of $\theta^2-\phi-1 = 0$, where ϕ is a root of $\phi^2-x = 0$. In this case θ is a root of $\theta^4-2\theta^2-x+1 = 0$, and the general result is proved by van der Waerden [1949, p. 107]. We say that a function is *elementary* if it can be expressed in terms of exponentials, logarithms and algebraic functions, such as $\log(1+\sqrt{(1+e^x)})$. We say that a function is *purely transcendental elementary* if no algebraic functions are needed in the representation of the function in terms of logarithms and exponentials, either directly, or via such techniques as writing $\sqrt{x}$ as $e^{(\log x)/2}$. We should note that trigonometric functions can be expressed in terms of exponentials, and inverse trigonometric functions in terms of logarithms, so that these classes are wider than they might seem.

Risch [1969] gives an algorithm for deciding if a purely transcendental elementary function has an elementary integral or not, and for finding the elementary integral if it exists. This algorithm was rapidly implemented [Moses, 1971] (though some of the supporting algorithms were not as general as they could have been). If this algorithm does not find an integral, then that is a proof that no elementary form for the integral exists.

More recent developments have included algorithms for the integration of algebraic functions [Davenport, 1981; Trager, 1984], and for the integration of functions which can be expressed as purely transcendental elementary functions of algebraic functions [Davenport, 1984] or n-th roots of purely transcendental algebraic functions [Trager, 1979]. These later algorithms have not been implemented in full generality, due to the great complexity of general algebraic functions, though the case of square roots was largely treated in [Davenport, 1981]. Again, these algorithms will, by their failing to give an integral, prove that an integral has no elementary closed form.

One should not under-estimate the complexities of these processes. Early parts of the theory of integration (including the well-known result that $\exp(-x^2)$ has no elementary integral) were due to Liouville [1835], and used to be included in standard Cours d'Analyse until about 1880, when they were replaced by easier and more "modern" material. For a long time it was said that only Risch and Moses understood Risch's paper, and even now there is only one widely-available implementation of that algorithm. The more recent develop- ments, involving the theory of algebraic functions, and hence algebraic geometry, are mathematically much more sophisticated. The great advantage of a computer implementation is that the user need not understand the internal mathematics, only the definition and use of integration.

More recently, Risch and Norman (unpublished see Davenport [1982] and the references cited there for a complete description) have produced a simplification of this process, which, while not quite complete, systematizes the method of integration by parts to the point where any integral of a purely transcendental elementary function can be reduced to closed form or to a "simple" non elementary integral. As Fitch [1981] has pointed out, this method can be applied to non-elementary functions as well, though then it will not *prove* that integrals have no closed form. This method is substantially simpler to program, and has been implemented in several computer algebra systems.

Why Have Closed Forms?

If one wishes to compute the numeric value of a single integral, then direct numeric evaluation has many advantages over a lengthy, and possibly fruitless, search for a closed form. The numeric methods are much more widely available than computer-based symbolic integration, and much less error-prone than manual integration, they cope with a much wider range of integrands, and are generally much faster. The situation is very different when one wishes to compute many integrals of the same form, i.e. many values of $F(y)$, as defined by

$$F(y) = \int_{a(y)}^{b(y)} f(x,y)\, dx ,$$

where, in general, y represents several parameters. Lyness [1983] opines that the first question that a numerical analyst should ask his customer is "How many integrals do you want?". The most obvious case of this, as Fateman [1981] has pointed out, is that of multiple integrals, where a single numeric answer can require the evaluation of thousands of one-dimensional integrals to construct the multi-dimensional answer. However, numeric multi-dimensional integration is a relatively well-understood problem, and the example that the author has seen more often is that of minimization of an integral.

As a simple example of this, consider

$$\min_{a\geq 0} \int_0^\infty (ax+a^{3/2})\, e^{-ax}\, dx = \min_{a\geq 0} \frac{a^{3/2}+1}{a} = 2^{2/3} .$$

The formal integration, and indeed the formal resolution of the entire problem, is trivial, but let us examine a naive approach to this problem. Since it is generally recognised that minimization is more efficient when derivatives are provided, we will try both with and without first derivatives.

	Integration	
	formal	numeric (D01AMF)
no (E04ABF)	0.0476	3.061
yes (E04BBF)	0.0160	4.374

These results are not very surprising, and point out a further advantage of symbolic integrals, viz. that the derivatives (with respect to other parameters) can be calculated almost immediately (in particular, if f is a rational function of $x_1,\ldots,x_n$, Baur & Strassen [1983] prove that the cost of computing f and all $\partial f/\partial x_i$ is at most three times that of computing f), whereas numeric integrals tend to require further integrations, each as difficult as the function itself. In this case, the extra expense is unjustified. The only other point worth noting is that a naive approach to this problem fails, since it finds a minimum at $a=\infty$, because, although the integral tends to infinity as a tends to ∞, the integrand is represented as zero by the computer throughout its range when a is large.

Let us now consider a more complex example, though still much simplified from real life, taken from a problem posed to the author and B. M. Trager a few years ago. The simplified problem is to calculate

$$\min_{\substack{0\leq\eta\leq 1\\ 0\leq\xi\leq 1}} I = \min_{\substack{0\leq\eta\leq 1\\ 0\leq\xi\leq 1}} \int_{y=\eta}^{y=1} \int_{x=0}^{x=1} f(x,y,\eta,\xi)\, dx\, dy$$

where f is defined as

$$\frac{\eta x\xi}{\eta^2(1+x^2\xi)+y^2-y\sqrt{(y^2+\eta^2(1+x^2\xi))}} - \frac{\xi(1-\xi)}{1-\eta} - \eta$$

(in the actual problem that was posed, f depended on more variables, and the minimum was required for several values of the extra variables). The existence of a closed form for the integral follows from the general theory of integration [Hardy, 1916, ch. 4] and the fact that the residues of f, when f is considered as a function of x, are independent of *y*, though the actual closed form is somewhat messy.

The timings obtained, both with and without the use of first derivatives, for this calculation are:

Derivatives:	no	no	yes	yes
	E04JBF	E04JBF	E04KDF	E04KDF
Integration:	formal	numeric	formal	numeric
		D01DAF		D01DAF/D01AHF
Accuracy (absolute)				
10^{-1}	0.1139	0.8287	.04741	0.8793
10^{-2}	0.1149	1.8287	.06020	1.5459
10^{-3}	0.1424	2.8287	.05992	1.7667
10^{-4}	0.1468	2.8287	.07231	4.3847
10^{-5}	0.1476	2.8287	.07253	5.8955
10^{-6}	0.1474	2.8287	.07247	8.1025
10^{-7}	0.1501	4.8287	.07293	10.0975
10^{-8}	0.1831	10.8287	.07249	10.9257
10^{-9}	0.1977	11.8287	.07231	12.9263
10^{-10}	0.2103	13.8287	.07235	18.6988
10^{-11}	0.4368	*	.07245	21.1160
10^{-12}	*	*	.07254	22.6353
10^{-13}	*	*	.09760	38.1690

An asterisk indicates that the minimization method was unable to produce an answer which it could guarantee to the stipulated accuracy. We note again that the methods using derivatives were much more successful in terms of returning an answer, though the use of numeric integration and derivatives is very slow. The apparent sharp jump in the last column between 10^{-3} and 10^{-4} is probably caused by the fact that, near the minimum, the expression for one partial derivative is obtained by cancelling nearly equal terms, viz.:

$$\frac{\partial I}{\partial \eta} = \int_{y=\eta}^{y=1} \int_{x=0}^{x=1} \frac{\partial f(x,y,\eta,\xi)}{\partial \eta}\, dx dy - \int_{x=0}^{x=1} f(x,\eta,\eta,\xi)\, dx ,$$

where the two integrals must be calculated to far greater precision than the final result (whether this is done symbolically or numerically, though if the integrals are calculated symbolically there is a greater chance of being able to rearrange the calculation so as to avoid the cancellation). Observation of the integration procedures at work (both the one-dimensional and two-dimensional integrals were evaluated with methods based on optimal extension of Gaussian quadrature [NAG, 1982]) showed that they did not change the evaluation strategy (i.e. the number of points) in the neighbourhood of the minimum. Had this not been the case, the minimisation would have been very much more difficult [Lyness, 1977].

The purpose of this section is to show that, although numeric integrals may be much easier to obtain (no integration system available to the author is capable of performing the above integration without assistance, though it is algorithmically solvable), they may not be adequate, and that, for reasons of accuracy or of computing cost, a closed form is sometimes preferable. In view of this requirement for integrals, however difficult, it is regrettable that some integration implementations (MACSYMA included) do not give certain integrals which have closed forms, believing that the closed forms are too complex. Even if no human wanted to look at them, they may still have value as input to a further numeric process.

How Many Integrals Have Closed Forms?

The question posed in the heading is clearly vital for this paper, since there is no point is exhorting readers to look for objects which almost never exist. However, it is very difficult to see how to make the question precise. In a purely statistical sense, almost all integrals (of functions more complicated than rational, or rationalizable, functions) do not have closed forms. This is analogous to the result that almost all polynomials are unfactorizable, and probably about as meaningless. Despite the fact that almost all polynomials are unfactorizable, the polynomials that one meets in everyday problems frequently factorize, and knowledge of the factorization is extremely useful.

Similarly, it is the author's contention that many (though certainly not all) of the integrals that are encountered in real problems have closed forms, either completely or for some of the layers of a multi-dimensional integral. This is a contention that is clearly very difficult to substantiate, but of the fifteen or so integrals referred to the author by colleagues (which are therefore not a representative cross-section, since both the "trivial" and the "obviously hopeless" will have been excluded), about one-third turned out to have closed forms, while another one-third could be improved by the use of integration theory, either by integrating in one of the many dimensions, or by reducing the complexity of the integrand. Among those completely integrated was a numerical analysis "practical question" posed to undergraduates by a colleague, and deliberately contrived so as to require, in his opinion, numerical methods.

Such evidence is clearly only anecdotal, but it is not clear how to gain more accurate figures. The author hopes that it may persuade some readers of the value of looking again at their integrals, and consulting an expert, be it a program or a human. The necessity of using a program in complex cases is brought out by the following example.

Waldvogel [1983], when looking for singular perturbation expansions in van der Pol's equation, was led to consider a family of functions defined by:

$$c_j = \int \frac{(1/3 \cdot s)^4 f_j}{s^2} ds \; ; \; v_j = \frac{s\, c_j}{(1/3 \cdot s)^3} \; ; \; u_j = \int \frac{c_j}{(1/3 \cdot s)^2} \; ;$$

where f_j is a polynomial function of v_i and u_i ($i \leq j-1$). Waldvogel asserted that the analytic theory of integration was of no use to him, since these functions did not have closed form integrals from c_4 on. Computer-based investigation (using the system of Fitch [1981]) showed that c_4 did in fact have a closed form (about 10 lines long), and furthermore that the next few terms also had closed forms. This led to further theoretical work, and a demonstration that all the functions involved did, in fact, have closed forms.

Insoluble Integrals

Despite the optimism of the previous section, it is clear that there are many functions that do not have closed form integrals. Has the theory of integration anything to offer for these? The answer is in the affirmative: the theory of integration is capable of providing "reduced forms" for non-elementary integrals. This reduced form may be simpler to integrate because the integrand is cheaper to evaluate, or it may have other advantages. As a simple example of this, we quote the reduction

$$\int \log(e^{-x^2}+1) \left[2e^{-x^2} + \frac{e^{-x^2}+1}{x^2} \right] dx = \frac{e^{-x^2}+1}{x} \log(e^{-x^2}+1) + \int 2e^{-x^2} dx ,$$

where the partial integration has removed the logarithmic term from the integrand (as well as the division).

The reduced form may also have fewer singularities than the original, and it is a general principle of numeric integration that singularities, even if outside the immediate area of interest, are one of the major difficulties (see the discussion in Acton [1970]). One obvious example of this is the integral

$$f(y) = \int_0^y \log \sin x \, dx ,$$

which is not expressible in closed form, and where the integrand has a singularity at the origin. An attempt at analytic integration yields the equivalent form

$$f(y) = y \log \sin y - \int_0^y x \frac{\cos x}{\sin x} dx ,$$

which is far more tractable, though care still needs to be taken evaluating the integrand near $x=0$. To illiustrate this, we calculated $f(y)$ to 11 decimal places for y = .01(.01).1, using D01AHF. The results were

	direct	transformed
time(secs)	3.751	0.201
evaluations	1721	70

We emphasize that this transformation can be made automatically, as part of an attempt to calculate a closed form for the integral by the methods mentioned above. In view of this, it is a pity that not all implementations are capable of returning such partial results from an unsuccessful attempt at analytic integration. Of course, such partial results are not always better than the input, though they will often be. One unsolved research problem, which could loosely be categorized as that of "expert systems for numerical analysis" is that of deciding whether the partial results are better, or more generally that of deciding which of two forms of an integral is numerically preferable. In practice, an experienced numerical analyst has very little doubt, but a naiver user may well be confused.

Another example in this area is that any integrand of the form $f(x,y)$, where f is a rational function and y^2 is a polynomial of degree 3 or 4 in x, can always be reduced [Hardy, 1916] to the form

$$g(x,y) + \int h(x,y)\,dx + c \int \frac{1}{y}\,dx\ ,$$

where g is another rational function, h is a function with only logarithmic singularities, and c is a constant. Normally (but see Davenport [1981] for some counter-examples), h is integrable as a sum of logarithms, but, even when it is not, it is possible to move the logarithmic singularities of h away from a finite region of interest. The last term is not analytically integrable (unless $c=0$), but the differential has no singularities (the integrand has square-root singularities, but these will disappear under the appropriate transformation), and thus is relatively easy to integrate numerically. Indeed, special subroutines are often written for this integral, which is a version of the incomplete elliptic integrals.

Cherry [1983] (see also Cherry & Caviness [1984]) has developed an algorithm that will take purely transcendental elementary functions and express their integrals, if possible, in terms not only of elementary functions, but also of error functions or of logarithmic and elliptic integrals. This is capable of producing results like:

$$\int \frac{\log(x)^2}{\log(x)^2 + 3\log(x) + 2}\,dx = \frac{-7\mathrm{li}(e^2x)}{e^2} + \frac{4\mathrm{li}(ex)}{e} + x\ ,$$

though whether or not this is actually a simplification from the numeric point of view depends on the ultimate aim of the integration.

Of course, there are many such simplifications that are currently beyond any theory of integration, such as the proof that the non-logarithmic part of any integral of the form

$$\int \frac{R(x)}{\sqrt{(ax^6+bx^4+cx^2+d)}}\,dx\ ,$$

where R is a rational function of x, can always be reduced to the sum of elliptic integrals [Hardy, 1916, pp. 50-51].

Intelligent Expressions for Integrals

There are often many ways of expressing the same numeric integral, some of which are far more efficient than others. NAG [1982, sub D01DAF] quotes the example of

$$\int_R (x+y)\, dxdy = \int_0^1 \int_0^{\sqrt{(1-y^2)}} (x+y)\, dxdy$$

(where R is the positive quadrant of the unit circle) which can be expressed in a more efficient form as

$$\int_0^1 \int_0^{\pi/2} r^2(\cos\theta + \sin\theta)\, d\theta\, dr \;.$$

To six decimal places, the first formulation takes 189 evaluations, while the second takes 89. Of course, the answer can easily be evaluated analytically as 2/3.

A more expensive variant of this problem was presented to the author some years ago (in his role as a general Duty Consultant, not as an integration specialist). The user was trying to integrate $f(x,y)$ over the box $0 \leq x,y \leq 1$, and the software used was complaining that it needed more than 255 points to evaluate the integrals. Investigation showed that the function f was $1/\sqrt{(x^2+y^2)}$ if $x^2+y^2 \leq 1$, and 0 otherwise -- a true case of trying to fit a round peg in a square hole. This discontinuous behaviour for each value of x was causing the integration routines enormous problems, whereas in fact the user was trying to integrate $1/r$ over the quadrant R. When this was pointed out to him, he said (incorrectly) that there was no routine for integrating over that area, so he had just enclosed his area in a box. Doing transforms was "all too difficult".

Such a user could have been helped by an "intelligent" numerical integration package, that was capable of looking at the form of the integrand, rather than merely calling a FORTRAN function to evaluate the integrand at selected points. While no such software currently exists, it is not beyond the reach of computer scientists and numerical analysts to implement such software, and the gain in user productivity, and experts' freedom from trivial enquiries, would be substantial.

Conclusions

The main conclusion that we draw from the above discussions is that, imperfect though it may be, the subject of integration in closed form can be helpful to the client with a problem to solve, even one whose answer is purely numeric. There is still much to be done by the integrators before such methods become commonplace, but integration algorithms and systems do exist, and they have proved useful, solving not only problems that were evidently within their reach, but also problems that were thought to be purely numerical.

This is not to say that formal integration is now, or is soon likely to be, at a level where is can be used completely automatically, as differentiation is [Rall, 1981]. If the integral has a closed form, then a FORTRAN program to evaluate it can be written automatically: if not, then some manual intervention is necessary. Whether the program produced is efficient or well-conditioned is another question that has to be considered, though some progress is being made on this front [Wang et al., 1984; Hulshof & van Hulzen, 1984].

Besides the obvious goals of greater theoretical progress in the field of integration and the development of an integration "expert system" combining both algebraic and numeric knowledge, more widespread diffusion of the existing systems is clearly necessary. The rapid change of microcomputer technology is of great help here. The machine that Hearn [1980] forecast is largely available, and indeed the author has just seen demonstrated a single-user, 4 megabyte, REDUCE machine for £8,000, which will therefore run the software described by Fitch [1981] very easily. Indeed, a small system running off floppy discs would cost about £3,000.

References

Acton, 1970

Acton, F.S., *Numerical Methods that Work.* Harper & Row, 1970.

Baur & Strassen, 1983

Baur, W., & Strassen, V., *The Complexity of Partial Derivatives.* Theor. Comp. Sci. 22(1983) pp. 317-330.

Cherry, 1983

Cherry, G.W., *Algorithms for Integrating Elementary Functions in Terms of Logarithmic Integrals and Error Functions.* Ph.D. Thesis, U. of Delaware, August 1983.

Cherry & Caviness, 1983

Cherry, G.W. & Caviness, B.F., *Integration in Finite Terms with Special Functions: A Progress Report.* Proc. EUROSAM 84 (Springer Lecture Notes in Computer Science 174) pp. 351-358.

Davenport, 1981

Davenport, J.H., *On the Integration of Algebraic Functions.* Springer Lecture Notes in Computer Science 102, Springer Verlag, 1981.

Davenport, 1982

Davenport, J.H., *On the Parallel Risch Algorithm (I).* Proc. EUROCAM 82

(Springer Lecture Notes in Computer Science 144), pp. 144–157.

Davenport, 1984

Davenport, J. H., *Intégration algorithmique des fonctions élémentairement transcendantes sur une courbe algébrique*. Annales de l'Institut Fourier **34**(1984) pp. 271–276.

Fateman, 1981

Fateman, R. J., *Computer Algebra and Numeric Integration*. Proc. SYMSAC 81, ACM, New York, pp. 228–232.

Fitch, 1981

Fitch, J. P., *User-based Integration Software*. Proc. SYMSAC 81, ACM, New York, pp. 245–248.

Hardy, 1916

Hardy, G. H., *The Integration of Functions of a Single Variable*, 2nd. ed. C.U.P., 1916.

Hearn, 1980

Hearn, A. C., *The Personal Algebra Machine*. Proc. IFIP 80, Elsevier North-Holland, 1980, pp. 621–628.

Hulshof & van Hulzen, 1984

Hulshof, B. J. A. & van Hulzen, J. A., *Automatic Error Cumulation Control*. Proc. EUROSAM 84 (Springer Lecture Notes in Computer Science 174) pp. 260– 271.

Liouville, 1835

Liouville, J., *Mémoire sur l'intégration d'une classe de fonctions transcendantes*. Crelle's J. **13**(1835) pp. 93–118.

Lovelace, 1844

Lovelace, A. A., *Sketch of the Analytic Engine Invented by Charles Babbage* (L. F. Menabrea, trans. A. A. Lovelace) Translator's Note A. Taylor's Scientific Memoirs **3**(1844). (The relevant passage is quoted by D. E. Knuth in *The Art of Computer Programming Vol. I*, Addison-Wesley, 1983, p. 1).

Lyness, 1977

Lyness, J. N., *An Interface Problem in Numerical Software*. Proc. Sixth Manitoba Conf. Numerical Methods and Computing (1976), Congr. Numerantium XVIII,

1977, pp. 251-263.

Lyness, 1983

Lyness, J. N., *When Not to Use an Automatic Quadrature Routine.* SIAM Review **25**(1983) pp. 63-87.

Moses, 1971

Moses, J., *Symbolic Integration, the Stormy Decade.* Comm. A.C.M. **14**(1971) pp. 548-560.

Nolan, 1953

Nolan, J., *Analytic Differentiation on a Digital Computer.* M.A. Thesis, M.I.T. Dept. of Math., May 1953.

NAG, 1982

The NAG Library Manual, Mark 9. Numerical Algorithms Group, Limited. Oxford, 1982.

[Norman & Moore, 1977]

Norman, A.C. & Moore, P.M.A., *Implementing the New Risch Integration Algorithm.* Proc. 4th. Int. Colloq. Adv. Computing Methods in Theor. Physics, Marseilles, 1977, pp. 99-110.

Rall, 1981

Rall, L.B., *Automatic Differentiation - Techniques and Examples.* Springer Lecture Notes in Computer Science 120, Springer Verlag, 1981.

Risch, 1969

Risch, R.H., *The Problem of Integration in Finite Terms.* Trans A.M.S. **139**(1969) pp. 167-189.

Trager, 1979

Trager, B.M., *Integration of Simple Radical Extensions.* Proc. EUROSAM 79 (Springer Lecture Notes in Computer Science 79) pp. 408-414.

Trager, 1984

Trager, B.M., *Integration of Algebraic Functions.* Ph.D. Thesis, M.I.T. Dept. of EE&CS, August 1984.

van der Waerden, 1949

van der Waerden, B. L., *Modern Algebra*, Frederick Ungar, New York, 1949.

Waldvogel, 1983

Waldvogel, J., *Private Communication*, June 3rd, 1983.

Wang et al., 1984

Wang, P. S., Chang, T. Y. P. & van Hulzen, J. A., *Code Generation and Optimization for Finite Element Analysis*. Proc. EUROSAM 84 (Springer Lecture Notes in Computer Science 174) pp. 237- 247.

International Scientific Symposium of IBM Germany
"ACCURATE SCIENTIFIC COMPUTATIONS"
March 12-14, 1985, Bad Neuenahr

COMPUTER ALGEBRA AND EXACT SOLUTIONS TO SYSTEMS OF POLYNOMIAL EQUATIONS

Dr. B. M. Trager, IBM Research Yorktown Heights

Abstract:

Systems of polynomial equations arise in many diverse areas of scientific endeavor. Questions of detecting over- or under-determined systems are very difficult using only numerical techniques. The desire to locate and isolate all the solutions of such a system as well as to handle systems with parameters led to the application of techniques from symbolic computation. The initial approach of classical elimination theory, while providing a theoretical solution, was only able to solve the most trivial problems in a practical amount of time. The recent development of a Grobner Basis Algorithm for polynomial ideals seems to provide hope for isolating all solutions for a much larger class of problems. This algorithm can provide a set of "triangular" generators for the polynomial ideal in analogy with the triangular matrix produced by Gaussian elimination for solving linear equations. By "backsolving" these triangular generators, all solutions to the original system can be found. Inconsistent systems are easily recognized by the presence of a non-zero constant among the generators and under-determined systems lack "Pivot" Polynomials for at least one of the variables. An under-determined system does not have a finite number of solutions, but algorithms for constructive primary decomposition have been developed which can separate the solution manifold into its irreducible components.

The potential of Grobner Bases for other computational problems is also being investigated. The problems of multivariate polynomial factorization and greatest common divisor computation have been shown to be simply reducible to Grobner Basis computations. The computational complexity of the Grobner Basis Algorithm is still under study, but recent computational improvements are enlarging the class of problems to which this powerful approach can be successfully applied.

International Scientific Symposium of IBM Germany
"ACCURATE SCIENTIFIC COMPUTATIONS"
March 12-14, 1985, Bad Neuenahr

THE EUCLIDEAN ALGORITHM FOR GAUSSIAN INTEGERS

Prof. G. E. Collins, University of Wisconsin

Abstract:

Studies of the behavior of a version of the Euclidean algorithm for the Gaussian integers, the nearest remainder algorithm, are reported. Modifications of this algorithm are described which result in a much faster version for Gaussian integers in the typical size range encountered in applications of computer algebra systems.

AN EFFICIENT STOCHASTIC METHOD FOR ROUND-OFF ERROR ANALYSIS

J. Vignes*
R. Alt**

*Professeur à l'Université P. et M. Curie de Paris,
4, Place Jussieu - 75230 - Paris Cedex 05 (France).
Conseiller Scientifique à l'Institut Français du Pétrole
4, Avenue de Bois-Préau - 92500 - Rueil-Malmaison (France)

**Maître de Conférence à l'Université P. et M. Curie de Paris
4, Place Jussieu- 75230 - Paris Cedex 05 (France).

ABSTRACT

This paper presents a survey of research results obtained by the authors and by their team, on the round-off error propagation and the accuracy of mathematical computations.

The efficiency of the Permutation-Perturbation method is shown:

i) For evaluating the accuracy of the exact finite method results,

ii) For breaking off the iterative processes using the optimum termination criterion and evaluating the accuracy of the results,

iii) For choosing the optimum integration step in approximate methods, such as the finite difference method.

The Permutation-Perturbation method may be also used to evaluate the data error propagation.

Examples are presented to illustrate the efficiency of the method.

1. INTRODUCTION

From the mathematical standpoint, numerical methods may be classified in three categories:

(*i*) *Finite Exact Methods* (FEM) which provide exact results in a finite number of computations. For instance Gaussian elimination and discrete linear mathematical transforms belong to this category.

(*ii*) *Iterative Methods* (IM) which provide exact results as the limit of a sequence. For instance Raphson-Newton's method for solving systems of nonlinear equations or any optimization method may be classified in this category.

(*iii*) *Approximative Methods* (AM) which provide only an approximation of the exact results. For instance any finite difference method belongs to this category.

From the computer science standpoint, when we perform these methods on a computer, several problems arise.

- For finite exact methods the results supplied by the computer always contain an error which is the consequence of the round-off error propagation in the floating-point arithmetic of the computer. For any of these methods the problem is: What is the accuracy of the results ?

- For iterative methods, the results supplied by the computer contain a double error, one resulting from the termination criterion of the iterative process, and the other resulting from round-off error propagation. Consequently for these methods there are two problems to be solved:

(*i*) How to break off the iterative process ?

(*ii*) What is the accuracy of the results ?

- For approximative methods the results supplied by the computer also contain a double error, one resulting from the approximation and the other resulting from round-off error propagation. Consequently a problem must be solved:

How to choose the step for minimizing the result of these errors?

This paper presents an efficient practical method for solving these problems.

2. FINITE EXACT METHODS (FEM)

Any FEM may be defined by

$$\text{procedure } (d, r, +, -, x, :, \text{ funct}) \tag{1}$$

where $d \subset \mathbb{R}$ is the set of data

$r \subset \mathbb{R}$ is the set of results

For the sake of simplicity, we assume that the result r is unique. Consequently: $r \in \mathbb{R}$.

+,-,x,:,funct are the arithmetic operators.

For performing this algebraïc procedure on a computer we write it, using a programming language, in a syntactic form equivalent to its algebraïc form, and we obtain a data-processing procedure defined by

$$\text{PRØCEDURE}(D, R, \oplus, \ominus, *, /, \text{FUNCT}) \tag{2}$$

where $D \subset \mathbb{F}$ is the set of data

$R \in \mathbb{F}$ is the result

$\oplus$, $\ominus$, $*$, $/$, FUNCT are the data-processing arithmetic operators. $\mathbb{F}$ is the set of values, represented in the computer, in the standardized floating-point-mode.

2.1. Permutation-Method

Since floating-point arithmetic does not satisfy the rules of algebra, such as the associativity of the addition, there is not one computer procedure image of the algebraïc procedure but rather a set S of C_{op} computer procedure images with C_{op} being the number of possible permutations of the arithmetic operators. Consequently, we have

$$\text{Card } S = C_{op} \tag{3}$$

2.2. Perturbation Method

Since each floating-point operator produces a round-off error, we must admit that for each arithmetic operator there are two results, one by lack, the other by excess, both of which legitimatly represent the exact result. Consequently, if the procedure contains k arithmetic operators, each computer procedure supplies 2^k results, which are all equally representative of the unique algebraïc result r.

2.3. Permutation-Perturbation Method

By applying Perturbation Method to each computer image, we obtain a set $\mathcal{R}$ of results $\{\mathcal{R}/R_i,\ R_i \in \mathbb{F}\}$, with each R_i representing as legitimatly as the other the unique algebraïc result r. We have

$$\text{Card } \mathcal{R} = 2^k C_{op} \tag{4}$$

2.4. Evaluating the accuracy on the result

Maillé has shown [10] that the mean value $\overline{R}$ of the elements of $\mathcal{R}$ is the best approximation of r, and that the number C of decimal significant figures on $\overline{R}$ is given by

$$\text{Card} = \log_{10} \frac{|\overline{R}|}{\delta} \tag{5}$$

with δ being the standard deviation of $\mathcal{R}$.

2.5. Practical Permutation-Perturbation Method

In practice, it is impossible to generate all the elements of $\mathcal{R}$. Therefore, we must find a subset of $\mathcal{R}$ which is representative of $\mathcal{R}$.

It has been shown [10] that it suffices to perform a computer procedure image three times by randomly permuting the linear operators and by randomly perturbing (add 0 or 1 at the lowest weight bit of the mantissa) the result of each arithmetic operator. So, with these three elements R_1, R_2, R_3, we compute the computer result $\overline{R}$ and the standard deviation δ, and with Equation (5) we obtain the number C of decimal significant figures in $\overline{R}$.

Random permutation-perturbation is performed with a Fortran Function PEPER. This function needs approximatively 30 statements. This is very easy to use in any algorithm. Then we have a practical tool, enabling us to evaluate the accuracy of the results of any finite exact method.

3. STOCHASTIC APPROACH FOR ROUND-OFF ERROR PROPAGATION

As developed in [13] any arithmetic operation defined by

$$z = x \omega y \tag{6}$$

with $x, y \in \mathbb{R}$ and $\omega \in [+, -, x, :]$

is performed on a computer as

$$Z = X \Omega Y \tag{7}$$

with $X, Y \in \mathbb{F}$ and $\Omega \in [\oplus, \ominus, *, /]$, $\oplus, \ominus, *, /$ being the computer arithmetic floating-point operators.

It is easy to see that the absolute error ε_Z defined by

$$\varepsilon_Z = Z - z = X \Omega Y - x \omega y \tag{8}$$

may be easily evaluated as shown below.

Generally x and y may not be coded exactly in the computer and we have:

$$\begin{aligned} X &= x(1+ \alpha_x) & \alpha_x &\in P(\alpha) \\ Y &= y(1+ \alpha_y) & \alpha_y &\in P(\alpha) \end{aligned} \tag{9}$$

α_x and α_y being the relative round-off error. These errors may be considered as random independant variables belonging to a statistical distribution P whose mean value $\overline{\alpha}$ and standard deviation σ depend on the computer used. For computers working in chopping or rounding normalized floating-point arithmetic with base 2 or 16 with n bits in the mantissa, the values $\overline{\alpha}$ and σ are given in [1].

Because of the round-off error when Z is computed, we obtain:

$$\begin{aligned} Z &= X \Omega Y = X \omega Y + \alpha_z(X \omega Y) \qquad \alpha_z \in P(\alpha) \\ Z &= (x+\alpha_x) \omega (y+\alpha_y y) + \alpha_z((x+\alpha_x x) \omega (y+\alpha_y y)) \end{aligned} \tag{10}$$

So the absolute error ε_Z is given by:

$$\varepsilon_Z = Z - z = (x+\alpha_x x) \omega (y+\alpha_y y) - (x \omega y) + \alpha_z((x+\alpha_x x) \omega (y+\alpha_y y)) \tag{11}$$

It is easy to see that, since each exact arithmetic operator is continued and derivable in $\mathbb{R}$, by neglecting the error of 2nd order the absolute error ε_Z or relative error ε_Z/Z of each floating-point arithmetic operator is an inner product of the α errors.

Indeed

$$\text{if} \quad \Omega = \begin{cases} + \\ - \end{cases} \quad \text{then} \quad \varepsilon_Z = x\,\alpha_x \pm y\,\alpha_y = (x \pm y)\alpha_z \qquad \alpha_x, \alpha_y, \alpha_z \in P(\alpha)$$

$$\text{if} \quad \Omega = \begin{cases} * \\ / \end{cases} \quad \text{then} \quad \varepsilon_Z = r\,\alpha_x \pm r\,\alpha_y + r\,\alpha_z \qquad \text{with} \quad r = x * y \quad \text{or} \quad r = x / y$$

The same is also true for the relative error.
So Z is defined by:

$$Z = z + \varepsilon_Z = z\,(1 + \frac{\varepsilon_Z}{z}) = z\,(1 + \beta_Z) \tag{13}$$

β_Z is the round-off error propagation which must be considered as a random variable. At each step computation the error in a result is always composed of two parts, one resulting from the round-off error propagation of the previous steps which is an inner product of two vectors, one being a random variables vector, the other parts results from the round-off error in this step which must be also considered as a random variable.

Consequently if we consider the algebraïc procedure defined by Eq.(1) requiring k arithmetic operations (assignments, +, -, x , :) and producting a single result r, the absolute error ε_a and the relative error ε_r in the result R is given by Eqs.(14):

$$\varepsilon_a = \sum_{i=1}^{i=k-1} g_i(d)\,\alpha_i + \alpha_k r \tag{14}$$

$$\varepsilon_r = \sum_{i=1}^{i=k-1} h_i(d)\,\alpha_i + \alpha_K \qquad \alpha_i, \alpha_k \in P(\alpha)$$

$g_i(d)$ and $h_i(d)$ being functionals of the data d.

Each α_i may generally be considered as an independant random variable, and consequently ε_a or ε_r which are inner products of α_i must be considered as normal random variables in agreement with the central limit theorem.

Now we assume that the algebraïc procedure is performed with all possible data, then the signs of functionnals $g_i(d)$ and $h_i(d)$ are generally randomly positive or negative. Consequently ε_r must be considered as a normal variable with a null mean.

4. PROOF BY SIMULATION ON A COMPUTER

As shown in section 2 using the permutation-perturbation method, the number C of significant decimal figures of any result provided by a computer is given by Eq.(5).

Maillé [10] has shown that, with three random elements R_i, $i = 1,2,3$ the exact value r belongs to the interval I_e defined by Ed.(15)

$$I_e = [\,\overline{R} - 2.48\,\delta,\quad C + 2.48\,\delta] \tag{15}$$

with 95% confidence, δ being the standard deviation of R_i.

Now let $r \in R$ be the exact result of any computation, and R^* a computer representation. By using simulation code a statistical analysis has been made on the difference DC defined by:

$$DC = |C - C_e| \tag{16}$$

C_e being the exact number of significant decimal figures of R^*. Indeed the simulation code provides normal random variables R_i with mean R^* and standard deviation σ.

So the exact number of significant decimal figures C_e is defined by:

$$C_e = \mathrm{Log}_{10} \frac{|R^*|}{\sigma} \tag{17}$$

The statistical analysis has been made with $R^* = 1.;10.;10^3;10^6$; and $\sigma = 1$.

By using Eq.(5) and Eq.(17), C and C_e are computed for each value of R^*, at first with 3 and next with 2 random variables R_i, and the corresponding DC is computed with Eq.(16). This absolute error DC is evaluated 10 000 times for each values of R^*. A mean average absolute error DCM and a standard deviation EDC of this absolute error are then evaluated for each value of R^*. Then, the number of values DC which are greater than a level ECC, choosen a priori, are counted. So the histogram of DC is obtained. The levels ECC chosen vary from 0.15 with a step of 0.15. With this method the confidence interval of C may be evaluated.

The results obtained with normal random variables with R^* as mean are given in Table 1. From the results, it appears immediately that, with a probability 95%, C is evaluated with Eq.(5) with an absolute error less than twice the value of the standard deviation.

So we can be sure that, if C is computed with 3 random results with Eq.(5), then the maximal absolute error on C is 0.44 with a confidence interval of 95% because the standard deviation of DC is 0.22 as shown in Table 1.

If C is computed with Eq.(5) with only two random results, then the maximal absolute error for C is 0.56 because the corresponding standard deviation of DC is 0.28. To show the robustness of the method, another simulation with the same code was performed with a uniform distribution on $[\frac{-\sqrt{12}}{2}, \frac{\sqrt{12}}{2}]$. The results obtained are given in Table 2. We can see that the results are practicaly the same as those obtained with normal distribution. We can then conclude that this method for evaluating the number of significant decimal figures for an any result is very efficient and in agreement with the theoretical results presented by Maillé [10]. Indeed from the result given in Table 1 we can see that the number C of exact significant decimal figures in a result computed with Eq.(5) from three random results R_i $i = 1,2,3$ belongs to the interval I_k defined by:

$$I_k = [C - 2\sigma, C + 2\sigma] \tag{18}$$

with 95% confidence, σ being the standard deviation of C.

Percent DC greater than ECC
Normal distribution

ECC	$R^* = 1.0$		$R^* = 10^2$		$R^* = 10^3$		$R^* = 10^6$	
	3 results	2 results	3 results	2 results	3 results	2 results	3 results	2 results
0.15	56.5	60.8	46.0	59.4	46.7	59.3	46.7	59.3
0.30	13.6	17.8	13.2	25.6	12.8	25.9	12.8	25.9
0.45	3.9	7.1	3.5	8.2	3.5	8.2	3.5	8.2
0.60	1.4	3.3	1.2	3.4	1.4	3.0	1.4	3.0
0.75	0.4	1.7	0.4	1.8	0.4	1.6	0.4	1.6
0.90	0.2	0.8	0.1	0.9	0.15	0.9	0.15	0.9
1.05	0.1	0.5	0.06	0.5	0.08	0.5	0.08	0.5
1.20	0.04	0.3	0.01	0.1	0.02	0.2	0.02	0.2
1.35	0.01	0.1	0.0	0.05	0.01	0.1	0.01	0.1
1.50	0.0	0.06	0.0	0.03	0.0	0.03	0.0	0.03
Standard deviation of C								
	0.22	0.27	0.21	0.28	0.21	0.28	0.21	0.28

<u>table 1</u>

Percent DC greater than ECC
Uniform distribution

ECC	R* = 1.0		R* = 10^2		R* = 10^3		R* = 10^6	
	3 results	2 results	3 results	2 results	3 results	2 results	3 results	2 results
0.15	51.9	57.6	35.5	60.3	36.3	61.3	35.7	61.8
0.30	7.7	11.4	4.1	14.8	4.0	14.9	3.5	14.1
0.45	1.6	3.8	1.1	3.2	1.4	3.5	1.1	3.5
0.60	0.6	1.9	0.4	1.5	0.5	1.7	0.4	1.8
0.75	0.2	0.7	0.08	0.8	0.2	0.8	0.2	1.0
0.90	0.03	0.4	0.04	0.4	0.05	0.4	0.09	0.5
1.05	0.01	0.2	0.03	0.2	0.02	0.2	0.03	0.2
1.20	0.01	0.1	0.01	0.1	0.0	0.08	0.0	0.1
1.35	0.0	0.04	0.01	0.08	0.0	0.07	0.0	0.03
1.50	0.0	0.04	0.01	0.02	0.0	0.04	0.0	0.01
	Standard deviation of C							
	0.20	0.24	0.15	0.22	0.16	0.22	0.15	0.22

table 2

5. APPLICATION OF THE PERMUTATION-PERTURBATION METHOD

5.1. Some elementary examples of functionals

The following examples are quoted from Kulisch [8].

5.1.1. The exact result for the computation of

$$R = 9x^4 - y^4 + 2y^2$$

for $x = 10864$ and $y = 18817$ is $R = 1$.

This expression has been computed on an IBM 3081 and the number of decimal significant digits on the value of $\overline{R}$ obtained, in using the permutation-perturbation method. The following results have been obtained:

- single precision 0 significant figure
- double precision 0 significant figure
- quadruple precision $R = 1.000...0$ with 36 significant figures

5.1.2. $P = 12192\,x^3 - 32257\,x^2 - 85344\,x + 225799$

compute P for $x = 2.645752$

On the same IBM 3081 computer the following results have been obtained:

- single precision 0 significant figure
- double precision $P = 3.052.... \times 10^{-8}$ with 3 significant figures
- quadruple precision $P = 3.056390553600... \times 10^{-8}$ with 30 significant figures.

5.2. Linear Algebra

The permutation-perturbation method has enabled us to very easily solve all problems encountered in linear algebra (Matrix singularity, accuracy of the solution of linear systems of equations, accuracy and improvement of eigenvalues, etc.). In linear algebra, it is very easy to perform the permutation-perturbation method. It suffices to apply the perturbations only to the elements of the matrices and vectors, while the permutations are applied to the colomns or the rows of the matrices.

We present here only the results concerning the detection of matrix singularity and the accuracy of the solution of the systems of linear equations.

5.2.1. Detection of matrix singularity

A matrix is numerically singular when the computed value of its determinant is either null or nonsignificant. Consequently when the number C of significant decimal figures in the computed value of the determinant of a matrix is less than one $(C < 1)$, then the matrix is numerically singular.

It is very easy to evaluate C with the permutation-perturbation method. The results concerning Hilbert's matrix H_n are given here:

$$H_n = [a_{ij}] \qquad a_{ij} = \frac{1}{i+j-1} \qquad i,j \in [1,n] \tag{19}$$

This matrix H_n is always mathematically regular, but it is very ill-conditioned and becomes numerically singular when n increases. Of course the numerical singularity depends on the precision of the computer arithmetic.

The results given in Table 3 show that Hilbert's matrix becomes numerically singular when $n = 12$ for a CDC 7600.

n	Δ^*	7600 CDC computer (p=48 bits)			Singularity S
		$\overline{\Delta}$	C^*	C	
2	8.3×10^{-2}	8.3×10^{-2}	13	13	
3	4.6×10^{-4}	4.6×10^{-4}	12	12	
4	1.6×10^{-7}	1.6×10^{-7}	11	11	
5	3.7×10^{-12}	3.7×10^{-12}	9	9	
6	5.3×10^{-18}	5.3×10^{-18}	8	8	
7	4.8×10^{-25}	4.8×10^{-25}	7	7	
8	2.7×10^{-33}	2.7×10^{-33}	5	5	
9	9.7×10^{-43}	9.7×10^{-43}	4	4	
10	2.1×10^{-53}	2.1×10^{-53}	3	3	
11	3.0×10^{-65}	2.8×10^{-65}	1	1	
12	2.6×10^{-78}	-4.5×10^{-78}	0	0	S
13	1.4×10^{-92}	-7.8×10^{-91}	0	0	S

Table 3

p is the number of bits in the mantissa

Δ^* is the exact value of the determinant

C^* is the exact number of significant decimal figures evaluated by comparing Δ and Δ^*.

5.2.2. The accuracy of the solution of a linear system

It is well known that when the matrix of a system of linear equations is ill-conditioned, the computed solution supplied by a computer does not have good accuracy.

But even if the matrix is well-conditioned we can obtain a misleading solution. This is the case for the system proposed by J.H. Wilkinson: $W_N X - B = 0$ with

$W_N =$

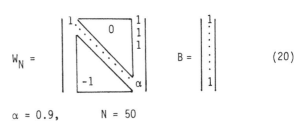

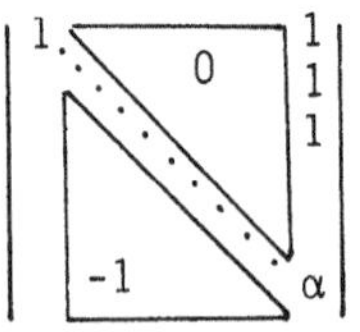

$$B = \begin{pmatrix} 1 \\ \vdots \\ 1 \end{pmatrix} \qquad (20)$$

$\alpha = 0.9, \qquad N = 50$

The exact solution of this system is

$$x_k = -2^{k-1}\,\frac{1-\alpha}{\Delta^*}\,, \qquad x_N = \frac{2^{N-1}}{\Delta^*}$$

with $\Delta^* = 2^{N-1} - 1 + \alpha$.

The use of the permutation perturbation method for evaluating the accuracy of the solution shows that the first eight unknowns are not significant ($C_i = 0$, $i = 1,\ldots,8$) as shown in Table 3.

Table 3

Variables	x_1 to x_8	x_9 to x_{11}	x_{12} to x_{14}	x_{15} to x_{18}	x_{19} to x_{23}	x_{24} to x_{25}	x_{26} to x_{28}	x_{29} to x_{33}	x_{34} to x_{36}	x_{37} to x_{39}	x_{40} to x_{43}	x_{44} to x_{47}	x_{48} to x_{49}	x_{50}
C_i	0	1	2	3	4	5	6	7	8	9	10	11	12	13

The permutation-perturbation method shows that it is impossible to solve this system by using Gaussian elimination.

5.3. Discrete mathematical transforms

Several papers have been published on round-off propagation in the Fast Fourier Transform (FFT) giving lower or upper bounds [7] as well as the mean square [1] of the overall round-off error propagation. But the permutation-perturbation method applied to any discrete mathematical transform (FFT, Walsh, Hadamard transforms, etc.) enables us to evaluate the local accuracy for each point of the transform. The corresponding Fortran code of the FFT has been published in [3].

Table 4 gives the results of the FFT (Cooley-Tuckey algorithm) of a complex function (1024 samples). The samples are randomly generated complex numbers which are uniformly distributed between the values -1 and +1. Since too large a table is required to represent 1024 values of the Fourier Transform, we present only a certain number of them chosen at random.

Table 4 shows the agreement between the number of significant decimal figures C determined by the Permutation-Perturbation method and the number of exact decimal figures obtained by a run in double precision.

Table 4

j	Real Part		Imaginary Part	
1	27.362054314926023855		- 22.540676373160984080	
	27.3620543149268	C = 14	- 22.5406763731612	C = 13
31	0.30436576039091963768		- 6.3199102026420999694	
	0.304365760389297	C = 11	- 6.31991020264309	C = 12
79	0.67643823378508592243		26.370038791154481990	
	0.676438233781973	C = 11	26.3700387911574	C = 13
275	- 9.3062710329200841270		10.360918417726311801	
	- 9.30627103292579	C = 12	10.3609184177295	C = 13
448	39.225546346328048287		- 8.5848579092036735106	
	39.2255463463464	C = 12	8.58485790920599	C = 12
526	20.315072867168419363		20.300843998858786857	
	20.3150728671692	C = 13	- 20.3008439988596	C = 13
528	1.0384628470167623661		25.399857391735793637	
	1.03846284701817	C = 12	25.3998573917352	C = 14
749	- 0.48928836241720576025O		2.1267963964204776780	
	- 0.489288362417199	C = 12	2.12679639642626	C = 12
799	6.8723607808165800158		0.081482780233098124808	
	6.87236078082236	C = 11	0.0814827802247082	C = 9
854	- 14.479042895715677390		- 0.029257041129125689884	
	- 14.4790428957155	C = 14	- 0.0292570411406814	C = 9
878	- 26.330955013083845979		- 0.27666192684760097446	
	- 26.3309550131019	C = 11	- 0.276661926841181	C = 11
962	0.71015027929282147140		0.86511479870731649062	
	0.710150279280430	C = 10	0.865114798709573	C = 11
1024	- 4.7055515901892891596		- 15.543975041433886085	
	- 4.70555159019762	C = 11	- 15.5439750414376	C = 13

5.4. Iterative methods

Let us consider an iterative method defined by the following sequence

$$X^{(0)}, X^{(0)}, X^{(2)}, \ldots, X^{(n)}, \ldots \qquad (21)$$

$X^{(j)}$ may be a scalar, a vector, a matrix.

When the method is convergent $X^{(n)} \to X_s$ when $n \to \infty$. X_s is the limit of the sequence.

When we implement an iterative method on a computer, we obtain a sequence

$$X^{(0)}, X^{(1)}, X^{(2)}, \ldots, X^{(n)} \qquad (22)$$

and two problems arise.

(*i*) How to break off the iterative process?

(*ii*) What is the accuracy of the result?

The permutation-perturbation method enables us to answer to these questions.

5.4.1. The standard termination criterion

The termination criterion generally chosen is

$$\|X^{(n)} - X^{(n-1)}\| \leq \varepsilon \quad \text{or} \quad \|X^{(n)} - X^{(n-1)}\| \leq \varepsilon \quad \|X^{(n)}\| \qquad (23)$$

with ε chosen arbitrarily.

If ε is chosen too large, then the iterative process is broken off too soon and the solution supplied is not satisfactory.

If ε is chosen too small, then a great number of iterations are performed without improving the accuracy of the solution. Sometimes, in this case, the round-off error propagation makes the iterative method divergent. Consequently the usual termination criteria are wrong.

5.4.2. The optimal termination criterion

The optimal termination criterion for iterative methods providing directly checkable results is given here.

Let us consider the example of an iterative method for solving a system of nonlinear equations

$$f_i(x_1, x_2, \ldots, x_n) = 0 \qquad i = 1, 2, \ldots, n \qquad (24)$$

From the mathematical standpoint, when the solution $x_1^*, \ldots, x_n^*$ is reached, we have:

$$f_i(x_1^*,\dots,x_n^*) = 0 \qquad i = 1,2,\dots,n \tag{25}$$

When we implement the iterative method on a computer, we have to solve the following system

$$F_i(X_1,X_2,\dots,X_n) \qquad i = 1,\dots,n \tag{26}$$

But, because of the round-off error propagation, even if we compute the functions with the X_i^*, $i = 1,2,\dots,n$ which are the computer images of the mathematical solution x_1^*, $i = 1,2,\dots,n$, we obtain:

$$F_i(X_i^*,\dots,X_n^*) = \rho_i \qquad i = 1,2,\dots,n \qquad \text{each } \rho_i \neq 0$$

is the residual due to the cumulative round-off error propagation which is not significant.

The optimal termination criterion is one which breaks off the iterative process when a satisfactory solution is reached. This means that the ρ_i, $i = 1,\dots,n$ are non significant, that is to say that the number C_i, $i = 1,2,\dots,n$ of significant decimal figures for each F_i, $i = 1,2,\dots,n$ is less than 1 ($C_i < 1$).

The optimal termination criterion at each iteration consists in using the permutation-perturbation method on each F_i, $i = 1,2,\dots,n$ and in computing the corresponding C_i and in breaking off the iterative process when $C_i < 1$; $i = 1,2,\dots,n$.

5.4.3. The accuracy of the solution

It is absolutely necessary to determine the accuracy of the solution obtained. To do this, we must solve the system three times, always using the permutation-perturbation method and the optimal termination criterion. Using equation (5) we easily determine the number of decimal significant figures for each X_i, $i = 1,2,\dots,n$, and obviously the accuracy.

Two examples of the use of the optimal termination criterion are given here.

5.4.4. Application of the optimal termination criterion

5.4.4.1. Resolution of a system of non-linear equations

Let us consider the following system of three non-linear equations:

$$\begin{cases} f_1 = x_1^2 + 2x_2 - e^{x_2} + x_3 + \frac{8}{9} - 10^{-12} = 0 \\ f_2 = x_1x_2 + e^{-x_2} + x_1x_3 - 1 - \frac{10^{-12}}{3} = 0 \\ f_3 = x_1^3 + \cos x_2 - x_3 - \frac{1}{27} + 10^{-12} - 1 = 0 \end{cases} \tag{27}$$

The exact solution of this system is:

$$x_1^* = \frac{1}{3} \quad x_2^* = 0 \qquad x_3^* = 10^{-12}$$

This system has been solved by Raphson-Newton's method on a CDC 7600 computer. We have used both the standard and the optimal termination criterion. The results are given in the tables 5 and 6. The initializations are in both cases $x_1^o = 0.3$, $x_2^o = 10^{-15}$, $x_3^o = 10^{-10}$.

$x_1^{(8)}$	$x_2^{(8)}$	$x_3^{(8)}$
0.333 333 333 333 345	$0.127\,669\,980\,869\,59 \times 10^{-13}$	$0.994\,001\,583\,581\,755 \times 10^{-12}$

table 5

Standard termination criterion defined by $|X^{(n)} - X^{(n-1)}| \leqslant \varepsilon$, $X^{(n)} \in F^3$. $\varepsilon = 10^{-14}$
Number of iterations : 8.

$x_1^{(5)}$	$x_2^{(5)}$	$x_3^{(5)}$	Number of exact digits in the results
0.333 333 333 333 357	0	$0.100\,108\,456\,305\,459 \times 10^{-11}$	$CX_1 = 13$ $CX_2 = 15$ $CX_3 = 3$

table 6

Optimal criterion defined by $C_f < 1$, $f \in F^3$. Number of iterations : 5.

Table 6 clearly shows that the optimal termination criterion has broken off the iterative process as soon as the solution has been reached and that the number CX_i of significant figures on X_i, $i = 1,2,3$, has been perfectly determined.

5.4.4.2. A stabilization of Bairstow's method

Bairstow's method can be used to solve the algebraic equations defined by:

$$P_n(x) = \sum_{i=0}^{n} a_i \, x^{n-i} = 0 \qquad a_i \in \mathbb{R} \qquad i = 1,2,\ldots,n \qquad (28)$$

This method consists in determining a second degree polynomial $x^2 - sx + p$ which exactly divides $P_n(x)$. By solving the second degree equation, we obtain two roots of $P_n(x)$. This process is done again on the $P_{n-2}(x)$, $P_{n-4}(x)$ etc.. plynomials, until all the n roots are determined.

We can write:

$$\begin{cases} P_n(x) = (x^2 - sx + p)\ P_{n-2}(x) + R(x) \\ P_{n-2}(x) = b_0\ x^{n-2} + b_1 x^{n-3} + \ldots + b_{n-2} \\ R(x) = b_{n-1}(x - s) + b_n \end{cases} \tag{29}$$

By polynomial identification we obtain:

$$b_j = a_j + sb_{j-1} - pb_{j-2} \tag{30}$$

The polynomial $x^2 - sx + p$ exactly divides $P_n(x)$ when we have:

$$\begin{cases} b_{n-1}(s,p) = 0 \\ b_n(s,p) = 0 \end{cases} \tag{31}$$

This system of two non-linear equations is solved by Raphson-Newton's method.

If we use the standard termination criterion to break off Raphson-Newton's method, Bairstow's method is unstable. But in using the optimal termination criterion, at each iteration we obtain the best values for s and p, and then Bairstow's method is stable.

Hereunder is the example of the solution of the following equation

$$a_1 x^6 + a_2 x^3 + a_3 x + a_4 = 0$$

with
$a_1 = 0.170522639876489 \times 10^{+18}$
$a_2 = 0.180900824489071 \times 10^{-4}$
$a_3 = 0.10113316962264 \times 10^{-20}$
$a_4 = 0.383826436411105 \times 10^{-26}$

By using the standard termination criterion we obtain on a CDC 7600 computer the following solution:

$x_1 = 0.472694333823987 \times 10^{-7}$
$x_2 = 0.610096708605860 \times 10^{-7}$
$x_3 = 0.229246106606858 \times 10^{-7} + i0.485339090111795 \times 10^{-7}$
$x_4 = 0.229246106606858 \times 10^{-7} - i0.485339090111795 \times 10^{-7}$
$x_5 = -0.339174524605003 \times 10^{-7} + i0.493663192841173 \times 10^{-7}$
$x_6 = -0.339174524605003 \times 10^{-7} - i0.493663192841173 \times 10^{-7}$

In these results, x_1 is obtained with 2 decimal significant figures,
x_3, x_4 are " " 1 " " "
x_2, x_5 and x_6 are unsignificant.

By using the optimal termination criterion we obtain on a CDC 7600 computer the following solution:

$$X_1 = -0.47207606833449 \times 10^{-7}$$
$$X_2 = 0.59747333966347 \times 10^{-7}$$
$$X_3 = 0.23604348981134 \times 10^{-7} + i0.41110574719182 \times 10^{-7}$$
$$X_4 = 0.23604348981134 \times 10^{-7} - i0.41110574719182 \times 10^{-7}$$
$$X_5 = -0.29874212547583 \times 10^{-7} - i0.51562637073114 \times 10^{-7}$$
$$X_6 = -0.29874212547483 \times 10^{-7} + i0.51562637053114 \times 10^{-7}$$

This solution is an approximation of the exact solution with 14 decimal significant figures on each x_i, $i = 1,\ldots,6$. This example shows clearly that Bairstow's method is stable when we use the optimal termination criterion.

6. APPROXIMATIVE METHODS

The numerical methods used to compute numerical derivatives and definite integrals and to solve ordinary differential equations or partial derivative equations are among what we call the approximative methods. When we carry them out on a computer we must assign a value to the discretizing step. And thus we must deal with two types of errors acting in opposite directions: as the discretizing step decreases, the method error decreases, while the computing error increases. In fact the discretizing step must be chosen so as to minimize the overall error.

Many studies have been undertaken by our research team to determine the optimal step, in the numerical derivatives computations: J. Dumontet [5] [6], in the numerical integration: J. Vergnes [11] [12] , and in the resolution of the ordinary differential equations: R. Alt [2].

We present here a summary of the results obtained by R. Alt for determining the optimal step used in one-step methods for solving the ordinary differential equations.

6.1. Introduction

In the numerical computation of an initial value problem solution, two different kinds of error may arise. First, the integration method is an approximation and provides a solution which is theoretically different from the exact solution. The difference is the so called truncation error. Second, the floating-point computer arithmetic introduces a round-off error into the numerical process.

Many of numerical experiments have shown that for one-step methods, the local relative round-off error either increases or remains constant as the step-size decreases. On the other hand, it is well know that if the method is convergent, the truncation error decreases with the step-size.

Consequently, given an initial value problem, a numerical one-step method and a digital computer, there exist an optimal value for each step in the computation of the solution, where the combined truncation error and round-off error is minimal, that is to say that there is at least one value of the step for which the accuracy obtained for the numerical solution is maximal. The upper bound of these values is called "the local optimal integration step-size" and the aim of the work exposed below is its computation. Therefore, the study can be divided in three parts. First, the round-off error is taken into account, and it is shown that the best and easiest way to obtain the most accurate result of a computation is by using the permutation-perturbation method of J. Vignes and M. La Porte [13]. A statistical formula for specifying the round-off error on the solution in the case of a linear differential system is also derived from the above method.

Second, the several ways of evaluation of the truncation error are investigated and a new and more efficient method is proposed.

Third, the round-off error and truncation errors having been evaluated, an algorithm for the computation of the local optimal integration step-size is given.

6.2. Estimation of the round-off error

There are three conventional ways of evaluating the round-off error in a numerical process: interval analysis, Miller's device using Wilkinson's ideas and the permutation-perturbation method of J. Vignes and M. La Porte. For our purpose, we choose this last method for its simplicity and its efficiency. It must be used in the case of a non linear differential system. Its necessity is shown in paragraph 6.2.1. However, when the system is linear it may be replaced by a simple statistical formula given in paragraph 6.2.2.

6.2.2. The need for round-off error estimation can be seen in the following example:

Consider the two dimension system given by Cash [4]:

$$\begin{cases} y_1' = 0.01 - (0.01 + y_1 + y_2)\,(y_1^2 + 1001\, y_1 + 1001) \\ y_2' = 0.01 - (0.01 + y_1 + y_2)\,(1 + y_2^2) \\ y_1(0) = y_2(0) = 0 \end{cases} \qquad (32)$$

The integration is performed on a CDC 7600 computer with the classical fourth order Runge-Kutta method with different step-size. The numbers of significant decimal figures CY1 and CY2 in the two components of the computed solution after the first step are given in the table 7.

	10^{-4}	10^{-6}	10^{-8}	10^{-10}	10^{-12}	10^{-14}	10^{-16}
CY1	14	14	14	14	14	14	14
CY2	13	11	9	7	5	3	0

table 7

The permutation-perturbation method shows clearly here that a high order integration method is of no use when the step-size is too small, because the round-off error can be much greater than the truncation error, as is obvious in the second component of the solution.

The idea behind the permutation-perturbation method can now be used to derive a statistical formula for estimating the round-off error in linear system. In fact, when the system is linear, and the integration method is of a Runge-Kutta type, the only operations performed in the numerical process are scalar products of floating point values. Therefore, it is possible to estimate the round-off error in the result with a simple formula.

6.2.2. A statistical formula for the round-off error on scalar products of floating point values

The results shown in [2] are given briefly here.

Let
$$q = \sum_{i=1}^{N} x_i\, y_i \qquad x_i, y_i \quad \mathbb{R} \tag{33}$$

be the exact result of a scalar product of real numbers, and let

$$Q = X_1 * Y_1 + X_2 * Y_2 + \dots + X_N * Y_N \tag{34}$$

be the floating point result. X_i and Y_i, $i = 1,\dots,N$, are the floating point representations of the exact real numbers x_i and y_i.

Now let

$$\rho = Q - q \tag{35}$$

If we assume that the round-off error are independant of one another, the following formula can be derived from those of La Porte and Vignes [9]

$$\overline{\rho} = \overline{\alpha}\, q \,\frac{N^2 + 7N - 2}{2N}$$

$$\overline{\rho}^2 = \overline{\alpha}^2 \left[\frac{3N^3 + 41N^2 + 134N - 72}{12N} q^2 + \frac{(N-2)(N^2 + 3N - 6)}{12N} s^2 \right] \qquad (36)$$

$$+ \sigma^2 \left[\frac{N+1}{3} q^2 + \frac{N^2 + 19N - 6}{6N} s^2 \right]$$

$$s^2 = \sum_{i=1}^{N} (x_i y_i)^2$$

where $\overline{\rho}$ and $\overline{\rho}^2$ are respectively the mean value and the quadratic mean value of the round-off error ρ.

Alt performed many numerical experiments which proved the rightness of these formulas.

6.3. Estimation of the truncation error

The idea of the new method proposed by Alt [2] and exposed here for estimating the local truncation error is different from the methods previously used. This method has been used by P.E. Zadunaisky [15] for another purpose. The method proposed by Alt may be expressed as follows.

Define the differential system as

$$\begin{cases} y' = f(x,y) \\ y(x_0) = y_0 \end{cases} \qquad (37)$$

and let y_k be the numerical solution computed with any one-step method at a point x_k.

Consider now the associate differential system

$$\begin{cases} z' = f(x,y) + R'(x) - f(x,R(x)) \\ z(x_k) = y_k \end{cases} \qquad (38)$$

where $R(x)$ is any differentiable function in the integration interval.

The exact solution of (38) is obviously:

$$z(x) = R(x) \qquad (39)$$

Thus, the truncation error in the numerical solution obtained for system (38) with the same one-step method is known. So, if the function $R(x)$ is sufficiently close to the exact solution $y(x)$, the known truncation error in (39) can be taken as a good approximation of the truncation error in system (37). It can be seen that $R(x)$ is an osculating function of $y(x)$ of the same order as in the one-step method and very good results have been obtained with rational functions. As an example the local truncation error in the integration of equation

$$\begin{cases} y' = 10\,y - 11\,e^{-x} \\ y(0) = 1 \end{cases} \qquad (40)$$

with the conventional fourth order Runge-Kutta method and a constant step $h = 0.01$, are estimated with the step halving method and "our associated problem" method. The results are presented in table 8.

x	Truncation error		
	Step halving method	Associated problem method	Exact
0.2	0.953952×10^{-9}	0.109679×10^{-8}	0.108714×10^{-8}
0.4	0.953916×10^{-9}	0.109674×10^{-8}	0.108709×10^{-8}
0.6	0.953528×10^{-9}	0.109628×10^{-8}	0.108662×10^{-8}
0.8	0.949861×10^{-9}	0.109210×10^{-8}	0.108246×10^{-8}
1.0	0.916865×10^{-9}	0.105602×10^{-8}	0.104480×10^{-8}

Table 8

6.4. Computation of the local optimal step-size

As the round-off error and truncation error are well estimated at each step of the integration process, it is now possible to get a good evaluation of the local optimal step-size which can be defined as the greatest step for which the truncation error and the round-off are equal. It is assumed that the computed truncation error cannot be smaller than the round-off error.

The algorithm for computing the local optimal step size can be described as below.

- The differential system of dimension p is:

$$y' = f(x,y) \qquad y \in \mathbb{R}^p \qquad (41)$$

- The one step method of order r is:

$$y_k = y_{k-1} + h\,\varphi(x_{k-1}, y_{k-1}, h) \qquad (42)$$

X_{k-1}, H, Y_{k-1} are the floating point approximations of x_{k-1}, h and y_{k-1}, Φ is an algorithm of computation of φ.

1°) For each j component of the system we compute:

$$Y_k^j \leftarrow Y_{k-1}^j + H * \Phi\,(X_{k-1}, Y_{k-1}, H) \qquad j = 1,2,\ldots,p \qquad (43)$$

2°) For each j component we compute the round-off error and the number of decimal significant figures of Y_k^j using the permutation-perturbation method or the statistical formula. C_k^j is the number of significant figures of Y_k^j.

3°) For each j we compute the truncation error and the number of significant digits τ_k^j of Y_k^j due to this error using the associated problem method.

4°) For each j, we compute the difference : $D_j = C_k^j - \tau_k^j$, $j = 1,\ldots,p$. (44)

5°) If one of the D_j is negative, H is too small then $H \leftarrow 2 * H$ and go to 1°). If all D_j are nonnegative, continue.

6°) We compute $DM = \underset{j=1,p}{\text{Max}}\ D_j$. (45)

7°) If $DM = 0$, the H is the optimal step-size. If DM is positive, the truncation error is two great, then set

$$H \leftarrow \text{Max}(H/2,\ H * 10^{-DM/(r+1)}) \qquad (46)$$

and go to 1°).

This algorithm has to be completed with some evident precautions to take in count an eventual constant solution for which $||Y_k - Y_{k-1}|| = 0$. It has been used with success in numerous experiments.

7. CONCLUSION

We have shown in this paper that the permutation-perturbation method is a very efficient tool

(*i*) to evaluate the round-off error propagation and determine the accuracy of the results provided by the exact finite methods.

(*ii*) to break off the iterative processes just when a satisfactory solution is reached and to determine the accuracy of the solution.

(*iii*) to evaluate the computing error in the approximative methods.

As the computers become increasingly more powerful, it is becoming easier to solve complex numerical problems requiring many computations. The round-off error propagation is obviously greater as the number of computations increases. Consequently, it is becoming more and more imperative to check the validity of the results provided by the computer.

We are led to believe that the permutation-perturbation method which has been used in many complex algorithms, and has never failed, is the right tool for the purpose.

REFERENCES

[1] R. Alt, Error Propagation in Fourier Transform, Math Comp. Sim. XX (1978) p.37-43.

[2] R. Alt, Un logiciel pour l'intégration optimale des systèmes différentiels, thèse d'Etat, Paris, 1980.

[3] P. Bois and J. Vignes, A software for evaluating local accuracy in the Fourier Transform, Math. Comp. Sim., XXII (1980), p.141-150.

[4] J.R. Cash, High order Methods for numerical integration of ordinary differential equations Numer. Math.30 (1979) p. 385-409.

[5] J. Dumontet, Algorithme de dérivation numérique. Etude théorique et mise en oeuvre sur ordinateur. Thèse de 3ème cycle, Paris, 1973.

[6] J. Dumontet et J. Vignes, Détermination du pas optimal dans le calcul des dérivées sur ordinateur. RAIRO, vol. 11 n°1 (1977).

[7] T. Kaneko and B. Liu, Accumulation of round-off error in FFT. J. of Ass. Comp. Mach, 17 (1970), p.637-654.

[8] U. Kulisch, R. Albrech, Grundlagen der Computer arithmetik, Springer Verlag, (1977).

[9] M. La Porte et J. Vignes, Evaluation de l'incertitude sur la solution d'un système linéaire. Numer. Math. 24 (1975), p.39-47.

[10] M. Maillé, Méthode d'Evaluation de la précision d'une mesure ou d'un calcul numérique. Rapport IP. Université Paris VI (1979).

[11] J. Vergnes, Détermination d'un pas optimum d'intégration pour la méthode de Simpson, Math. Comp. Sim. XXII (1980), p.177-188.

[12] J. Vergnes and J. Dumontet, Finding an optimal partition for a numerical integration using the trapezoïdal rule. Math. Comp. Sim. XXI (1975), p.231-241.

[13] J. Vignes, New Methods for Evaluating the validity of the results of mathematical computations. Math. Comp. Sim. XX (1978), pp.227-249.

[14] J.H. Wilkinson, Rouding-Error in Algebraïc Processes, Prentice Hall, (1963)

[15] P.E. Zadunaisky, On the Estimation of errors in the numerical integration of ordinary differential equations. Numer., Math., 27 (1976), p.21-39.

Vol. 192: Automata on Infinite Words. Proceedings, 1984. Edited by M. Nivat and D. Perrin. V, 216 pages.1985.

Vol. 193: Logics of Programs. Proceedings, 1985. Edited by R. Parikh. VI, 424 pages. 1985.

Vol. 194: Automata, Languages and Programming. Proceedings, 1985. Edited by W. Brauer. IX, 520 pages. 1985.

Vol. 195: H.J. Stüttgen, A Hierarchical Associative Processing System. XII, 273 pages. 1985.

Vol. 196: Advances in Cryptology. Proceedings of CRYPTO '84. Edited by G.R. Blakley and D. Chaum. IX, 491 pages.1985.

Vol. 197: Seminar on Concurrency. Proceedings, 1984. Edited by S.D. Brookes, A. W. Roscoe and G. Winskel. X, 523 pages. 1985.

Vol. 198: A. Businger, PORTAL Language Description. VIII, 186 pages. 1985.

Vol. 199: Fundamentals of Computation Theory. Proceedings, 1985. Edited by L. Budach. XII, 533 pages. 1985.

Vol. 200: J.L.A. van de Snepscheut, Trace Theory and VLSI Design. VI, 0–140 pages. 1985.

Vol. 201: Functional Programming Languages and Computer Architecture. Proceedings, 1985. Edited by J.-P. Jouannaud. VI, 413 pages. 1985.

Vol. 202: Rewriting Techniques and Applications. Edited by J.-P. Jouannaud. VI, 441 pages. 1985.

Vol. 203: EUROCAL '85. Proceedings Vol. 1, 1985. Edited by B. Buchberger. V, 233 pages. 1985.

Vol. 204: EUROCAL '85. Proceedings Vol. 2, 1985. Edited by B. F. Caviness. XVI, 650 pages. 1985.

Vol. 205: P. Klint, A Study in String Processing Languages. VIII, 165 pages. 1985.

Vol. 206: Foundations of Software Technology and Theoretical Computer Science. Proceedings, 1985. Edited by S.N. Maheshwari. IX, 522 pages. 1985.

Vol. 207: The Analysis of Concurrent Systems. Proceedings, 1983. Edited by B.T. Denvir, W.T. Harwood, M.I. Jackson and M.J. Wray. VII, 398 pages. 1985.

Vol. 208: Computation Theory. Proceedings, 1984. Edited by A. Skowron. VII, 397 pages. 1985.

Vol. 209: Advances in Cryptology. Proceedings of EUROCRYPT '84. Edited by T. Beth, N. Cot and I. Ingemarsson. VII, 491 pages. 1985.

Vol. 210: STACS 86. Proceedings, 1986. Edited by B. Monien and G. Vidal-Naquet. IX, 368 pages. 1986.

Vol. 211: U. Schöning, Complexity and Structure. V, 99 pages. 1986.

Vol. 212: Interval Mathematics 1985. Proceedings, 1985. Edited by K. Nickel. VI, 227 pages. 1986.

Vol. 213: ESOP 86. Proceedings, 1986. Edited by B. Robinet and R. Wilhelm. VI, 374 pages. 1986.

Vol. 214: CAAP '86. 11th Colloquium on Trees in Algebra and Programming. Proceedings, 1986. Edited by P. Franchi-Zannettacci. VI, 306 pages. 1986.

Vol. 215: Mathematical Methods of Specification and Synthesis of Software Systems '85. Proceedings, 1985. Edited by W. Bibel and K.P. Jantke. 245 pages. 1986.

Vol. 216: Ch. Fernstrom, I. Kruzela, B. Svensson, LUCAS Associative Array Processor: Design, Programming and Application Studies. XII, 323 pages. 1986.

Vol. 217: Programs as Data Objects. Proceedings, 1985. Edited by H. Ganzinger and N.D. Jones. X, 324 pages. 1986.

Vol. 218: Advances in Cryptology – CRYPTO '85. Proceedings, 1985. Edited by H.C. Williams. X, 548 pages. 1986.

Vol. 219: Advances in Cryptology – EUROCRYPT '85. Proceedings, 1985. Edited by F. Pichler. IX, 281 pages. 1986.

Vol. 220: RIMS Symposia on Software Science and Engineering II. Proceedings, 1983 and 1984. Edited by E. Goto, K. Araki and T. Yuasa. XI, 323 pages. 1986.

Vol. 221: Logic Programming '85. Proceedings, 1985. Edited by E. Wada. IX, 311 pages. 1986.

Vol. 222: Advances in Petri Nets 1985. Edited by G. Rozenberg. VI, 498 pages. 1986.

Vol. 223: Structure in Complexity Theory. Proceedings, 1986. Edited by A.L. Selman. VI, 401 pages. 1986.

Vol. 224: Current Trends in Concurrency. Overviews and Tutorials. Edited by J.W. de Bakker, W.-P. de Roever and G. Rozenberg. XII, 716 pages. 1986.

Vol. 225: Third International Conference on Logic Programming. Proceedings, 1986. Edited by E. Shapiro. IX, 720 pages. 1986.

Vol. 226: Automata, Languages and Programming. Proceedings, 1986. Edited by L. Kott. IX, 474 pages. 1986.

Vol. 227: VLSI Algorithms and Architectures – AWOC 86. Proceedings, 1986. Edited by F. Makedon, K. Mehlhorn, T. Papatheodorou and P. Spirakis. VIII, 328 pages. 1986.

Vol. 228: Applied Algebra, Algorithmics and Error-Correcting Codes. AAECC-2. Proceedings, 1984. Edited by A. Poli. VI, 265 pages. 1986.

Vol. 229: Algebraic Algorithms and Error-Correcting Codes. AAECC-3. Proceedings, 1985. Edited by J. Calmet. VII, 416 pages. 1986.

Vol. 230: 8th International Conference on Automated Deduction. Proceedings, 1986. Edited by J.H. Siekmann. X, 708 pages. 1986.

Vol. 231: R. Hausser, NEWCAT: Parsing Natural Language Using Left-Associative Grammar. II, 540 pages. 1986.

Vol. 232: Fundamentals of Artificial Intelligence. Edited by W. Bibel and Ph. Jorrand. VII, 313 pages. 1986.

Vol. 233: Mathematical Foundations of Computer Science 1986. Proceedings, 1986. Edited by J. Gruska, B. Rovan and J. Wiedermann. IX, 650 pages. 1986.

Vol. 234: Concepts in User Interfaces: A Reference Model for Command and Response Languages. By Members of IFIP Working Group 2.7. Edited by D. Beech. X, 116 pages. 1986.

Vol. 235: Accurate Scientific Computations. Proceedings, 1985. Edited by W.L. Miranker and R.A. Toupin. XIII, 205 pages. 1986.